GREEK AGRICULTURE IN A CHANGING INTERNATIONAL ENVIRONMENT

Greek Agriculture in a Changing International Environment

DIMITRIS DAMIANOS
EFTHALIA DIMARA
KATHARINA HASSAPOYANNES
DIMITRIS SKURAS

Routledge
Taylor & Francis Group

LONDON AND NEW YORK

First published 1998 by Ashgate Publishing

Reissued 2018 by Routledge
2 Park Square, Milton Park, Abingdon, Oxon OX14 4RN
711 Third Avenue, New York, NY 10017, USA

Routledge is an imprint of the Taylor & Francis Group, an informa business

Publisher's Note
The publisher has gone to great lengths to ensure the quality of this reprint but points out that some imperfections in the original copies may be apparent.

Disclaimer
The publisher has made every effort to trace copyright holders and welcomes correspondence from those they have been unable to contact.

A Library of Congress record exists under LC control number: 98073757

ISBN 13: 978-1-138-31919-6 (hbk)
ISBN 13: 978-1-138-31920-2 (pbk)
ISBN 13: 978-0-429-45405-9 (ebk)

Contents

Figures, maps and tables

Preface

This book is important for a number of reasons, and the English-reading world will gain from it. The most significant reason is that Greece is now a member of the EU, having joined in 1981. Adding Spain and Portugal to the Union in the round of enlargement that followed, in 1986, brought total Mediterranean membership of the EU to four member states (without including France which is at least partly Mediterranean). The four Mediterranean members now account for 43 per cent of crops, 37 per cent of grass, 25 per cent of woods in their one third of the EU agricultural area (Stanners and Bourdeau, 1995), and for one sixth of [EU] Mediterranean holdings. The average size of land holdings in Greece, which is slightly larger than 4 hectares, just under half of the group's average, emphasises the country's acute problems relating to rural development.

Its present position as a significant participant in European agricultural trade, as custodian of a substantial European heritage, as a member of the European Union and with its agriculture playing an important role in the Greek economy further adds to its significance.

As the book brings out with clarity, the world economy is now globalised and it is therefore becoming ever more vital to know about all parts of this vast network if any single state, or even group of states, is to participate in it effectively. Greek agriculture contributes a substantial share of national GDP and of exports. There are thus many parties who will need to know more about the economic shape of its agriculture in constructing and managing their own trade, agricultural and economic policies. For Greece to find its appropriate place, in the EU and the world, it will have to negotiate with many other states who will be more easily persuaded the better informed they are. Greece will, along with all other states, face a series of extended agricultural negotiations over the coming decade which will determine its contribution to the world trade well into the new century. Many participants in those discussions will need to know about Mediterranean agriculture, its problems and potentials, in some detail as the talks proceed. Beyond that immediate requirement a number of other issues will become contentious elements on the world policy agenda.

The Mediterranean itself is a virtually closed sea with which a number of other states have a frontier. Each of them is likely to lose from mismanagement of Greek agriculture and to gain as it approaches more sustainable development. Wider still, there are many recognised environmental pollutants to which Greek agriculture may add or from which it may subtract over coming decades, depending upon its developmental path. For example, fertilisers used in Greek agriculture, as in other parts of the world economy, cause soil degradation and pollution of water resources. There is potential for reducing environmentally hazardous effects by using sensible land use policies, policies which are more likely to be promoted from a better informed administration both in Athens and in Brussels.

Finally, Mediterranean states have many problems in common, a major one of which is how to modernise their agricultural sectors within the constraints of sustainable rural development. Whilst no state has the single key to these problems, all can learn from the policies pursued by others and for that reason this book will be of value.

With such an array of issues and problems bearing on the agriculture of EU member states, it is a pleasure to welcome a well constructed analysis of the position of Greek agriculture, its problems and prospects in a dynamic world economy.

Martin Whitby
Emeritus Professor of Countryside Management
The University of Newcastle upon Tyne

Reference

Stanners, D. and Bourdeau, P. (1995), *Europe's Environment: the Dobris Assessment: Statistical Assessment* Earthscan

Introduction

This book aims at identifying the scope of actions that need to be undertaken in order for Greek agriculture to survive the pressures from an increasingly competitive world. Two facts underscore this need. First, agriculture constitutes an important sector not only to Greece's economy, but also to the country's socio-cultural structure. Second, a series of international developments lead to several changes: to trade liberalization, to the expansion of existing markets and to the emergence of new ones, to improved market access, to changes in consumer priorities and to the awareness of environmental issues. In the meantime, the growth in information and communications systems amplify the impact that movements toward international economic integration have on economic and environmental sectors of various countries. Given these circumstances, the Greek agricultural sector and the related institutions should undergo major transformation. Well coordinated and concerted actions in various directions have to be undertaken in order for agriculture to contribute to the socio-economic well being of the countryside, consumers, and producers alike. The sector has to redefine its role regarding nature and small-scale family holdings, public health and the environment, rural development and the "grass-roots" approach, marketing and third country markets. The challenge is that these goals should be achieved under a system of reduced price support and no export subsidies.

In the 1980s, worldwide tensions in agricultural markets brought attention to the distortions that domestic agricultural policies of industrialized countries effect onto the world markets. Whether the result of realization of the world market distortions or of increasing budgetary pressures and of accumulating structural surpluses stemming from the continuation of price support policies in several industrialized states, a tendency toward trade liberalization and a switch to more competitive world markets of agricultural products began to be formed. The international environment was ready to consider the reform of agricultural policies.

In 1986 the last round of GATT's (General Agreement on Tariffs and Trade) multilateral trade negotiations began, and in 1987, OECD (Organization for Economic Cooperation and Development) declared the Ministerial Principles which called for improvements in the market orientation of agricultural sectors through concerted reductions in the support of agriculture and in the protection of agricultural markets. The Ministerial Principles aimed at the convergence of domestic prices of agricultural products to their respective world levels. Special weight was given to the reduction of measures that distort the world markets, by obstructing the transmission of world price movements to domestic producers. Transparency became a priority criterion in policy making, since it allowed for better planning of the distribution of benefits and prevented the leakage of benefits to other industries.

Five years later, the European Communities (EC) adopted Commissioner MacSharry's proposals for reform of the Common Agricultural Policy (the CAP). These proposals ensued under the pressure both of EC budgetary constraints and of the United States, a significant trade partner of the EC which orchestrated the latest GATT's negotiations. The new CAP brought EC prices closer to world market levels, introduced direct income aids, and implemented a set-aside scheme, thus reducing the support linked to production levels.

In 1994, the ratification of the Uruguay Round Agreement (URA) signaled a turning point. The Agreement, being in alignment with the 1987 OECD Ministerial Principles, accentuated the importance of policy transparency, market orientation and trade liberalization. It imposed the tariffication and the reduction of the heavily distorting non-tariff measures, restricted the disguised forms of protection, but exempted from reduction decoupled forms of support. The URA committed all contracting parties to reductions in domestic support, progressive liberalization of imports and restrained use of export subsidies.

At the same time, the EU acquired three new member states (Austria, Finland and Sweden). Moreover, in view of further enlargement toward Central and Eastern European Countries (CEECs) and Cyprus, the EU provided trade concessions to the prospective members.

Given the importance of agriculture to the candidate Member States, their accession to the EU was expected to have major implications for the future of EU's Common Agricultural and Structural Policies. In December 1995, the Community launched the "Agricultural Strategy Paper" which advocated further development of the CAP along the lines established by the 1992 reform. The Commission acknowledged the need for integrating market, rural development and environmental policies. The selection of appropriate instruments for this resolution was the next step.

In July 1997, the Commission proposed new reforms for confronting the conditions at the beginning of the 21st century. The need for further economic integration and institutional reform, in view of the imminent EU enlargement toward the CEECs and Cyprus, and the challenges of market internationalization required major changes relating to the Structural Funds, the CAP, the Enlargement and the Budgetary Framework. The Commission's proposals for reform were presented in the report "Agenda 2000: For a Stronger and Wider Europe".

All these international changes are expected to have a major impact on the agricultural sector and the rural society of Greece. More transparent and less distorting agricultural policies target the groups in need of assistance better and provide support, without introducing or accentuating market imbalances, financial burdens and income inequities. Improved market operation yields benefits in the long run, raises employment, reduces inefficiencies and inequities, and prevents overexploitation of natural resources and the environment. Realization of this prospect, however, necessitates structural adjustments by the economies involved.

Increasing world competition will force conventional agriculture to become more capital intensive, thus adding to the problems of rural unemployment and skewed distribution of income. Thus, an integrated approach toward sustainable rural development needs to be adopted. Conventional agricultural practices should target young farmers, whereas supplementary rural development policies should aim at absorbing the shocks in agriculture. The latter include the adoption of alternative farming systems, incentives for the settlement of young farmers, women, and specialized professionals in rural areas, and the effective use of Community programs.

The potential benefits expected to accrue from trade and market liberalization will be easily eliminated, should the necessary adjustments not take place. Non-adjusted countries will suffer the consequences of their structural weaknesses being exposed to the rest of the world and of the world's more competitive and stronger economic performance. Increasingly higher rates of unemployment, socio-economic instability and social disruption will be eventually experienced by the countries lagging in adjustment. Policies providing financial support without encouraging the necessary structural adjustments will only provide temporary relief and aggravate the situation in the long-run.

Greece avails itself for examining the impediments to agricultural development and for inquiring the potential of agriculture to contribute to rural sustainable development and social cohesion. Given the importance of the rural sector to Greece, rural sustainable development cannot be detached from the country's goal for socio-economic prosperity.

Agriculture is an important sector to the Greek economy. A significant part of the economically active population is fully or partly involved either in agricultural activities or those complementary to agriculture in the rural areas of the country. In certain regions, this percentage amounts to 30% of the economically active population. It follows that Greek agriculture has a significant role to play in keeping the socio-economic fabric of rural areas alive and in providing social services. Agriculture is one of the activities that are necessary for keeping the rural population from migrating to urban centers. Moreover, the adoption of extensive agricultural systems in mountainous areas can function as an economic activity that simultaneously safeguards the environment and protects the rural landscape, providing valuable services to society.

Crop and livestock production contribute, on average, 13% to the country's Gross Domestic Product. The country is self-sufficient in major crop products. Agricultural exports, mainly of crop products, account for 30% of the country's total exports and constitute a major source of foreign exchange revenues. Agricultural imports, consisting mainly of livestock and dairy products, account for 19% of the country's total imports. The value of agricultural imports, however, rises faster than the value of agricultural exports, resulting in negative trade balance for the sector.

Besides the importance of the primary sector in itself, agriculture is strongly linked to the food industry, one of the most dynamic sectors of the Greek economy. Agriculture's importance stems not only from its prominence in measurements of economic performance, but also in its potential for playing a key role to sustainable development. At the same time, the sector faces a series of impediments in attempting to reach this goal. Most of these impediments are inherent in the country's geoclimatic conditions and institutions.

Despite Greece's obligations toward the European Union to abide by the "Acquis Communautaire" and to conform with the CAP provisions, there is still room for designing a national strategy for sustainable agricultural and rural development. This is not only a possibility, but also the only way of successfully meeting current and future international challenges. This book builds up a framework of integrated policies that could confront unemployment, desertification of rural areas, the challenges of increasing competition, environmental degradation, and the irrational use of productive resources.

In order to pursue this goal, the book is organized so as:

- to present the socio-economic evolution of the agricultural sector in Greece;
- to review the agricultural, macroeconomic, rural and regional policies at the national or Community level, to analyze their impact on the development of the sector and to attempt to identify the factors responsible for limited effectiveness, for inefficiencies as well as for success;
- to summarize the latest international developments in policy making, trade, market operation and consumer preferences, to assess their implications for agriculture and rural communities and to inquire about the alternatives for surviving the challenges; and
- to develop a framework for an integrated strategy and to specify the actions necessary at the national, at the Community and at the inter-state level, in order to optimize the consequences of future international developments.

More specifically, the first chapter presents the evolution of key indicators for the agricultural sector in Greece, such as employment, income, trade, investment, farm structure and use of resources. The first chapter also focuses on the evolution of agricultural production patterns, identifies the factors that effected the changes, and assesses the adjustments in final production, in the value and productivity of intermediate consumption, in total factor productivity and in the demand for agricultural products.

Chapter two assesses the role of agriculture in the rural environment and reviews the role that institutions and farmers had on the dissemination of policies in Greek rural areas.

The third chapter provides an explanation for the performance of agriculture and for the situation in rural areas, presented in chapters one and two. The impacts of fiscal, monetary and exchange-rate policies on Greek agriculture are thoroughly assessed. Special attention is paid to the effects of the currency's overvaluation on agricultural trade and the cost of capital.

The fourth chapter assesses the impact that the country's accession to the EU had on the performance of the agricultural sector. A set of national objectives guided sectoral, regional and macroeconomic policy making.

From this set of factors those that contributed to policy success and those that restrained the sector's development and adaptation to the changing European and international environment are identified in chapter five.

The sixth chapter places Greek agriculture in the international economic environment and reveals the opportunities and constraints for rural development in Greece presented by changes in the CAP, the European Union enlargement and the international agreements on trade. It is proposed that under these developments, the agricultural sector of the country should be restructured along six distinct dimensions, five of which are dealt with in this chapter. An economic dimension, aiming at developing the competitive part of the sector; an ecological dimension, aiming at conserving and preserving the environment; a welfare dimension, aiming at minimizing inequalities in income distribution; a socio-cultural dimension, aiming at reinforcing local partnership and synergies; and a political dimension, aiming at the reconstructing institutions and raising public participation in policy making at local and regional levels. The sixth dimension of strengthening international economic relations and cooperation is examined in chapter four.

The seventh chapter delineates the essential elements of a framework for an integrated strategy for rural development in Greece. Sustainable rural development is viewed as a multi-axes pursuit, aiming at building a strong economic base for rural areas and at improving living conditions. Specific policies aiming at promoting job creation, enhancing factor mobility, protecting the environment, improving land management, advancing dissemination of information and education among the rural population, and restructuring cooperatives are considered. Reference is made to the need toward developing a basis for mutually beneficial economic cooperation, especially with the CEECs.

Special acknowledgments are extended to Ms. Amalia Melis for the relentless and adept editing of this book and for her stamina during our collaboration; to the services of the Ministry of Agriculture, and especially to Mr. D. Bourdaras for his generous and invaluable contribution to our edification on agricultural policy matters; to Professor Martin Whitby for taking precious time to write the preface for this book.

As regards our own roles, chapters one and two were written by E. Dimara and D. Skuras, chapter three and seven by K. Hassapoyannes, chapters four and five by D. Damianos, while chapters six and the conclusions were prepared by K. Hassapoyannes and D. Damianos.

Dimitri Damianos
Efthalia Dimara
Katharina Hassapoyannes
Dimitris Skuras

1 The evolution of the agricultural sector

A physical and historical background

Greece consists of a mainland and islands, that account for 81% and 19% respectively of its 132 thousand square kilometers. The islands number 217 inhabited islands and the country's coastline extends to more than 15 thousand kilometers, forming a remarkable series of bays and headlands. Almost 80% of mainland Greece is mountainous due to the southernmost extension of the Balkan peninsula. On the western side of Greece, the Epirus mountains continue the Alpine-Dinaric fold system and form the mountainous backbone of the country. Most mountain summits have an altitude of over 2,000 meters, that result in very large altitude differences. From the eastern borders of the country in Thrace to Northern Greece and down to the southern part of the island of Crete the central mountain range is crossed by smaller or larger river and torrent basins. Level areas in valleys or plains are close to the sea and mountain foothills and occupy almost 20% of the country's area. The four main rivers of the country, namely Evros, Nestos, Strymonas and Axios originate either from the Former Yugoslavian Republics or Bulgaria and their water management is the subject of transnational control and agreement.

Greece has a typical Mediterranean summer-drought climate with a strong maritime influence, as no part of the country is more than 90 kilometers away from the sea. During winter the weather is mild especially on the islands and along the west coast. Frosts are rare in the west and south and occur for less than 30 days on the coastal plains of Macedonia and Thrace. Spring is continental and of a short duration. During summer, mean temperatures on the lowlands rise to 26° or 27° C and sometimes heat exceeds 40° C. Autumn has an average temperature of 23° C in September with sunshine. The mean level

7

of annual precipitation is around 707 mm that is not evenly distributed among either places or time. Precipitation is regular in Western Greece on the Ionian side with a mean level twice higher than the national average. On the contrary, in areas of the Aegean the mean annual precipitation is almost half the national average. However, the largest and most intensively cultivated valleys in the east of mainland Greece are in need of irrigation. The time variation and uneven seasonal distribution of precipitation are a major constraint to agricultural development. Summer rain ranges from 7% to 46% of the annual precipitation level.

Rapid soil erosion has followed the clearance of the vegetative cover on steep slopes and hilly areas exposed to summer drought, heavy autumn and winter rains and the effects of more than 1,000 torrents. Evident erosion may be observed over almost 40 thousand square kilometers or almost 30% of the country's area. Nowadays, a decisive contribution to soil erosion is made by the numerous wild land and range land fires, over-grazing and incompatible land use. Soils are either thin and poor or ill-drained and only about 30% of the country's area is suitable for cultivation, while almost 40% is pastoral land. Forests occupy almost 18% of Greece and the major part of this is loosely spaced pine woods while the few tall forests surviving are on the most inaccessible and rainiest mountains of the country.

The modern Greek state became autonomous in 1827 following an independence war and revolution declared in 1821 against Turkey. Subsequent boundary changes included the annexation of the Ionian Islands in 1864, Thessaly and Arta in 1881, Macedonia, Epirus, Crete, Thrace and the Eastern Aegean Islands in 1912-13 and the Dodecanese Islands in 1947. The Independence War ascribed a vast amount of land to the newly created Greek state from the annexed regions, that was recognized internationally as national land. For 50 years, the state resisted pressure for a general distribution of the national land, but dissatisfaction among the large number of landless peasants, small tenants and subsistence farmers was increasing. An extensive land distribution was introduced in 1871 under which every adult Greek could apply for a plot of national land. Until 1911, almost 320,400 ha had been distributed to 387,137 beneficiaries with an average size of 1 ha for arable land and 0.3 ha for plantations and an overall average size of 0.82 ha. It is estimated that about 80% of the rural population became land owners of some kind as a result of the various land distribution schemes in this period (Anastasiades, 1911; Stefanides, 1948; Vernicos, 1973).

In the period 1881-1917, large estates emerged and dominated the prosperous regions of Thessaly, Macedonia and Western Thrace. In 1921-22 a war broke out with Turkey and under the Treaty of Lausanne in 1923, the two countries agreed to an exchange of populations. Nearly 1,625,000 refugees were accepted by Greece. This was a critical push and a decisive factor leading to the urgent implementation of an extensive land reform that affected 25% of the productive land including forests and chaparrals, and 36% of the agricultural utilized land (Alivisatos, 1932; Simonide, 1923). From the land reform almost 319,000 families received 2 million ha of land including ranges and forests. Distributed land usually consisted of 4 to 6 plots and, in some cases 10 to 18 plots, at different distances and directions from the village (Thompson, 1963). The structure of Greek agriculture today has been decisively affected by the long process that established the family farm as the core of agricultural production in the period 1821-1917.

After the Second World War and the severe civil war that followed, economic life was disrupted and many areas, especially Northern Greece, lost most of their resident population. Agricultural production was significantly lower than prior to war levels and most of the agricultural infrastructure was destroyed. In 1948, economic reconstruction started under U.S. aid, the greatest part of which was absorbed by agriculture. Large drainage and irrigation plans were designed and executed in the first 20 years after the war. Two of the most important drainage schemes reclaimed 20,000 ha of lake floor in Copais near the city of Athens and 40,000 ha of swampy lowlands on both sides of the city of Thessalonica. Large drainage and irrigation schemes combined with power plants were put into operation at the largest rivers of the country. In the same period, the Greek government accepted agricultural machinery, including 50,000 tractors, from the United Nations Relief and Works Agency (UNRWA) (Beckinsale and Beckinsale, 1975). By 1959, the country was, for the first time after WWII, self-sufficient in cereals. Agricultural production largely increased after 1960, and was assisted by the spread of cultivation such as cotton, tobacco and citrus and fruit plantations. The adoption of improved cultivation techniques accompanied by extensive fertilizer and pesticide use improved production. Vast credit provisions and protective price policies helped the situation. However, the post war period was marked by an enormous rural exodus involving almost one and a half million farmers because of surplus labor. Almost 60% of them immigrated to western European countries, mainly Germany, while the rest immigrated to Athens and other major urban centers (Carter, 1968; Dicks, 1967; Wagstaff, 1968).

From the early 1960s on and until the country's accession to the European Economic Community in 1981, national agricultural policy adopted marketing measures along with price supports that aimed to reduce existing differences between the various classes of the population and between regions. These measures were supplemented by other actions designed to raise the low standard of living of the rural population. The seriousness of farm structure problems was confronted by structural policies that considered an increase in the average size of farms and the consolidation of fragmented parcels of land.

A re-orientation of agricultural production was supported by encouraging the creation of large-scale industrial orchards, the adoption of genetically improved varieties, the introduction of new cultivation such as sugar beet and fodder plants as support to the livestock sector and the stabilization or even reduction of cereals.

The role of agriculture in the economy

This section provides a perspective of the agricultural sector's importance to the Greek economy by reviewing the developments in employment, income, trade and capital formation in agriculture and evaluates the performance of agribusiness.

Labor force in agriculture

The richest source of information on the workforce as a whole, is the decennial census of population that collects information from every household and communal establishment. The census of population records the "economically active population" including all those working for pay or profit in the week prior to the census together with those available for work at the time of the census. The economically active are subdivided into two groups, those in and those out of employment. The census of population provides an age, educational, social and occupational grouping of the labor force. The second most useful source on labor force in Greek agriculture is the decennial census of agriculture that collects information from every farming household and records the age and occupational characteristics of the household's head and his/her family. Other sources of information, that do not cover the entire population, are the regularly conducted farm structure surveys and the manpower survey. The former are conducted by the National Statistical Service of Greece (NSSG) under a European Union framework and the main results are made available by Eurostat, the European Union's statistical service, while the latter is irregularly conducted by NSSG and has a small sample coverage (about 1.5% of the actual economically active population). Labor force in agriculture has substantially decreased since 1960 but still accounts for the largest part of the total economically active population of the country. Almost half the economically active population in 1961 agriculture accounted for almost 16.7% in 1991 (figure 1.1). This decrease is due to the extensive rural out migration and the associated abandonment of agricultural activities in the period up to the early 1980s.

Economically active population

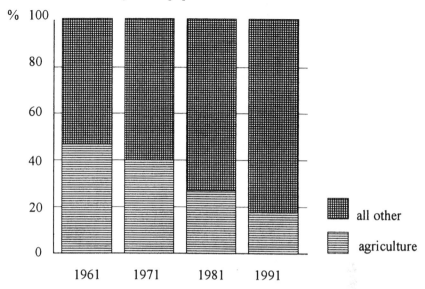

Economically active population

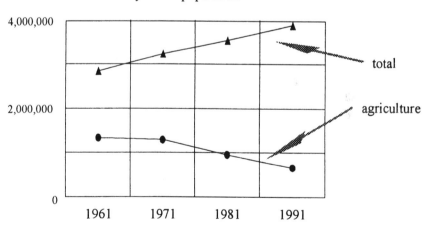

Figure 1.1: **The evolution of the economically active population in agriculture, relative to the total economically active population, 1961-91**

Source: *Censuses of population,* 1961-91, Athens, NSSG

However large the decline, the share of agriculture in the economically active population remains high compared to those of other European Union (EU) Member States. The family farm is the core of agricultural production and this is evident from a classification of the economically active population by professional status (figure 1.2). It is evident that the number of the "self-employed" has substantially increased its share among those economically active in the primary sector, while the share of the "family members" has decreased. It is interesting to note that the ratio of "family members" to "self-employed" has decreased from 0.8 to almost 0.35. This is again due to the rural exodus and the withdrawal of family members from farming activities due to the mechanization of the production process and the small size of the holdings. This change indicates a changing pattern of work allocation in the farm household and is associated with the changing character of Greek agriculture from numerous small family farms to fewer one-member farms. Thus, the meaning of family farm in Greek agriculture changes under technological developments that allow fewer members of the farm household to sustain the same levels of production and release labor from the agricultural sector.

The number of agricultural workers has substantially decreased in absolute numbers as would have been expected, but retains almost the same share of the economically active population in agriculture. Over time, the nature of salaried (paid) work in agriculture has changed from permanent full-time to seasonal employment. Demand for seasonal work is at its peak during the period that the major agricultural products are collected, namely cotton, tobacco, fruits, citrus fruits and olives. Wage rates for agricultural workers have been estimated to be much lower than in other EU Member States (Melas and Delis, 1981) and are frequently paid in product, especially in olive oil. Since 1990 many illegal immigrants from Albania work in agriculture with greatly reduced wage rates. Unfortunately, there are no official figures available estimating the participation of immigrants in the rural labor force but part of the increase in the agricultural production of the most labor intensive cultivation in recent years should be attributed to this fact. Finally, it should be noted that the only professional category that has increased in absolute numbers is that of "employees". The proportion of employees, employing permanent workers, who reside in urban and semi-urban areas is continuously rising and in 1991 was recorded to be around 27%. This is due to the growth of agricultural production enterprises, especially in the livestock sector, around major urban centers.

The main characteristics of the Greek workforce in agriculture are an age structure "top heavy" with the elderly, low skills and low level of education, unequal sex participation and extensive pluriactivity (Kasimis and Papadopoulos, 1997; 1994). An examination of the pyramid for the

12

economically active population in agriculture reveals the unusually high proportion of the elderly.

The effect of the high outmigration rates in the period 1965-75 may be seen in the sex pyramids for 1991 (figure 1.3). Almost 50% of the economically active in agriculture are over 50 years of age in both sexes, while almost 7.5% are over 65 years of age.

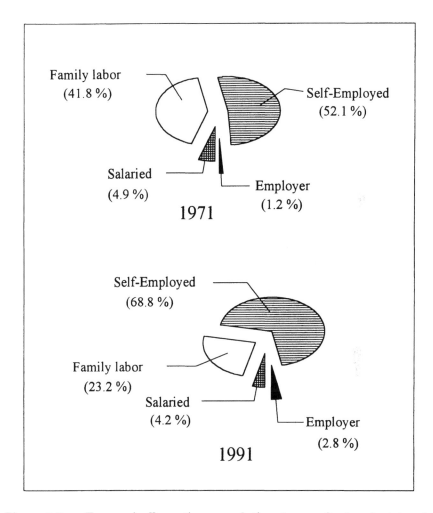

Figure 1.2: **Economically active population by professional status in agriculture**

Source: *Censuses of population,* 1971, 1991, Athens, NSSG

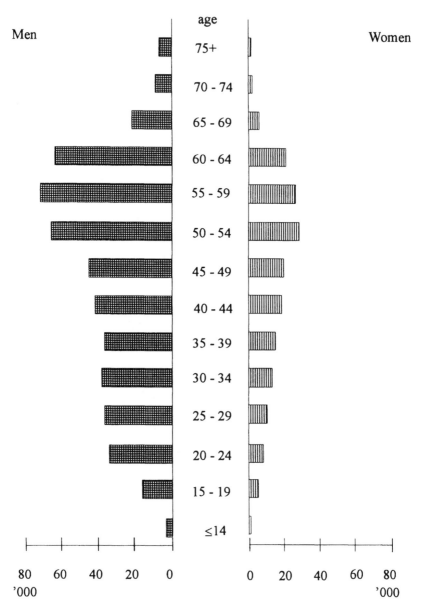

Figure 1.3: **Economically active population in agriculture by sex and age, 1991**

Source: *Census of population,* 1991, Athens, NSSG

14

A more gloomy and alarming picture is revealed when the age structure alone in relation to the size of the farm holding is considered. Almost 55% of the 65 years old and over farm owners, cultivate farms below 2 ha in size and only 4% cultivate large farms of over 10 ha in size (table 1.1). In all other age classes the corresponding percentages are not over 46 or below 8.5. It is clear that farmers over 65 years old enter a business contraction, survive and retain subsistence farms. If the same age classes are related to a classification of the days in a year worked on a farm, it is revealed that almost 55% of farmers over 65 years of age work for less than 75 days per year on the farm and only 11% work for more than 225 days.

Table 1.1
Percentage distribution of age by farm size

Age group	Farm size (hectares)				
	≤1.9	2-4.9	5-9.9	10+	Total
≤24	43.7	29.7	15.2	11.4	100
25-34	44.8	28.8	15.3	11.1	100
35-44	46.1	29.4	14.8	9.7	100
45-54	40.8	31.4	16.8	11.0	100
55-64	43.5	32.3	15.7	8.5	100
65+	54.5	30.4	11.0	4.1	100

Source: *Census of agriculture*, 1991, Athens, NSSG

The fact that older farmers remain active in agriculture is closely related to the attitudes toward the farm occupation and the retirement policy in the Greek agricultural sector. Since 1962, Greek farmers have been covered for retirement purposes by the Agricultural Insurance Organization that granted pensions on an age basis. The pension is awarded when the farmer reaches the age of 65 and is financed by the general state revenue. It is very low in value, approximating $120 per month in December 1996. Pensions do not constitute an independent self-sustaining viable income by any means and do not serve any agricultural structures policy such as the aim to rejuvenate agricultural population and release land to younger farmers. The early retirement pension provided by European Union Regulation 2079/92 (EEC, 1992) introduced higher pensions for those farmers between 55 and 65 years of age with the purpose of either withdrawing from production or passing the land over to younger farmers. This structural measure has not been extensively adopted, although it was expected to cover 50,000 farmers by 1997 and release 250,000 ha of land (Commission, 1996).

The education level of Greek farmers, compared to that of other sectors of the economy is very low. More than 95% of those economically active in agriculture have not or have finished elementary school only. The same holds true for agricultural training. The latest available survey of farm structures carried out in 1993, reports that less than 1% of the farm household heads have taken any kind of formal agricultural training and only 440 of them have taken a two-year agricultural training course. The overwhelming majority of Greek farmers relies heavily on practical experience. Education is a critical determinant of the quality of human capital on the farm and plays an important role concerning farmers' adaptation to the continuously changing economic environment, external to the farm. Empirical findings in many European and North American studies suggest that producers depend on their skills and creativity in finding and designing marketing systems outside the conventional set, or suggesting that educated farmers have better information processing abilities and are able to use sophisticated management techniques. Education is related to the receptiveness of the farmer. It is argued that educated farmers are likely to be adopters of new varieties or new agricultural technology while non-educated farmers are risk averse. Trained farmers are more dynamic in the adoption of new farming techniques and products, because they can either process information or value experience gained through work and are therefore more skillful.

Considering risky decisions, educated or trained farmers respond to policy signals and are likely to be "gamblers" whereas non-educated farmers are likely to be "satisfiers". The age and education of the farming population determine the rate and pace of innovation adoption and diffusion, and the success of dissemination and demonstration programs in agriculture and the operation of extension services. Greek and EU policies have attempted to increase the level of agricultural training among Greek farmers, through special training programs without much success. Special effort has been directed to the training of young farmers supported by the European Social Fund (ESF) and the European Agricultural Guidance and Guarantee Fund (EAGGF). It is expected that the level of education will increase in the future, taking into account the gradual withdrawal of older farmers and the continuation of the training schemes.

Multiple job holdings by farmers is a widespread phenomenon in Greek agriculture. Off-farm employment has been conceptualized in terms of "farm diversification", "other gainful activities" (OGA), "part-time farming", "pluriactivity" and "multiple job holding" (MJH). In this body of work the latter term has been adopted. It has been observed that the resort to main or exclusive employment outside the farm seems to be the result of deteriorating economic conditions for small farm households that no longer secure their social reproduction and economic viability (Damianos et al., 1991). This

16

development is historically rooted in the rural exodus started in the early 1950s, continued until the late 1960s and provided the initial growth of alternative employment opportunities outside the sector. Search for employment outside the farm involves heads and other members of the household. MJH of any kind among household heads is quite significant but seems to decrease in Greece while MJH with the main occupation outside the farm increased in the 1980s and seems to fall in the 1990s (table 1.2). MJH by any farm household member is extensive. Almost 42% of all farm households have at least one member with an occupation off farm and almost 36% have at least one member with a main or exclusive occupation off farm (table 1.3). These figures are lower for mountainous and less-favored areas due to restricted employment opportunities. MJH is also differentiated according to the farmer's age and the size of his farm household. Approximately 60% of the total number of household heads of ages 30 to 45 years maintain principle gainful activities off the farm. For small farmers, MJH takes the form of dependent labor in industry, tourism and construction while larger farmers direct their participation in MJH toward trade, services and other self-employed businesses (Moisidis, 1985).

A fourfold typology of MJH in Greece was developed according to the agricultural to non-agricultural transition stage (Damianos et al., 1991). Households with their main source of income coming outside agriculture have either completed their transition to the non-agricultural sectors or are at the last stage before the final exodus. The second type of households, called "worker-peasant" households, secure their reproduction mostly outside the farm, their agricultural employment has only a complementary role. They have a farm size usually smaller than 3 ha and are characterized by salary labor in industry, construction and tourism. The third type is considered as "farm households" as their holding area is usually above the national average and non-farm employment is of a secondary, complementary role and has a "diffused" character meant to contribute to further modernization and expansion of the farm business. The fourth type refers to the "farm business households", not in the sense of "agribusiness" but because the farm size and the technology used are good indicators of a process of accumulation. Off-farm employment in this group is unimportant and is not a necessity but rather a choice in the sphere of expanding their agricultural business in trading and processing agricultural produce.

MJH in Greece has been viewed as a survival strategy by which farm resources are reemployed into an off-farm occupation in response to falling agricultural incomes (Damianos and Skuras, 1996a). Off-farm employment is chosen as a survival strategy by low income farmers with large families and low agricultural debt. A very important role in choosing MJH as a survival strategy is played by the farmer's perception and information concerning the local and regional labor market in the other sectors of the economy.

Table 1.2
Multiple job holding (MJH) by farm household heads (%)

Year	MJH of any kind	MJH with a main occupation outside the farm
1977	37.1	21.8
1985	34.4	27.5
1993	29.6	24.2

Source: *Farm structure survey*, 1977-93, Athens, NSSG

Table 1.3
Percentage types of multiple job holding (MJH) in Greece, 1993

Type of MJH	Type of area		
	All	Mountainous	Less-Favored
MJH by any household member:			
of any kind	42.1	39.0	41.5
with a main occupation outside the farm	35.8	32.2	35.3
MJH by the household's head:			
of any kind	29.6	27.0	27.9
with a main occupation outside the farm	24.2	21.1	22.6

Source: *Farm structure survey*, 1993, Athens, NSSG

Such information is available to the farmer, among other sources, from members of the family living on the farm but working away from it. The extent of off-farm work carried out by family members is an indicator of the amount of non-agricultural information received by the farmer. The decision to participate in off-farm activities is influenced by human capital characteristics, and the property income of the farm household (Daouli and Demoussis, 1995). Off-farm labor supply is influenced by non-agricultural wage rates, the household's property income and the economic conditions prevailing in the local labor market (Daouli and Demoussis, 1995). Regional differences observed in MJH rates are positively related to off-farm work opportunities and negatively related to farm opportunities (push and pull factors). Research results support a push-pull hypothesis and indicate the relevant significance of push-pull factors under different regional socio-economic contexts (Efstratoglou-Todoulou, 1990).

Taking into account the extent of MJH in Greek agriculture, the number of people actually employed in agricultural activities should not be exclusively used as an employment indicator in Greece. A better picture of employment in Greek agriculture may be gained if either the time devoted to agricultural activities as a measure of actual work devoted to agriculture instead of the number of people involved were examined. More than 60% of farm household heads spend less than 149 work days a year on their farm, while only 21.5% work more than 225 days (figure 1.4). The fact that 60% of household heads work less than 149 days per year on the farm, does not necessarily mean that those farmers are pluriactive. An analysis of this information by age class reveals that almost 58% of these farmers are older than 55 years of age and thus may have entered a phase of business contraction without any other employment off the farm (figure 1.4). This hypothesis is supported by the low share of the over than 65 years age class, in the group of the more than 225 days of farm work per year.

It is important to note that only 21.5% of all farm household heads work full-time on their farm, i.e., for more than 225 days per year (figure 1.4). It is also important to approach employment in agriculture through a measure of work irrespective of the number of people involved in it, in order to reflect the role of part-time and seasonal work. A useful measure is the Annual Work Units (AWUs) which is equivalent to the time worked by one person employed full-time in agricultural activities on a holding over a whole year. Available data for Greece distinguishes between family AWUs i.e., the holder and members of his/her family working on the holding, and non-family AWUs i.e., paid workers not belonging to the holder's family. The two added together constitute the total AWUs. The volume of family and total labor input in Greek agriculture in AWUs has sharply decreased and in 1995 accounted for almost 70% of the corresponding 1980 level. The rate of decline, however, was not as high as in the other EU member countries and thus, the share of total Greek labor in total EU labor increased by more than 1%, that is from 8.7% in 1980 to more than 9.8% in 1993 (figure 1.5).

There is a strong regional dimension concerning the distribution of various labor characteristics over the country's territory. The regional character of Greek agriculture is revealed by many key agricultural variables that are highly differentiated with a clear trend to concentrate or locate in the different regions of the country. According to European Union territorial classification, the country is divided into 51 prefectures corresponding to the NUTSIII level of spatial disaggregation and 13 administrative regions corresponding to the

NUTSII level. Exposition to the regional distribution of the main agricultural economic features in Greece will be carried out at the NUTSII regional level according to the latest Survey on the farm structure survey carried out in 1993.

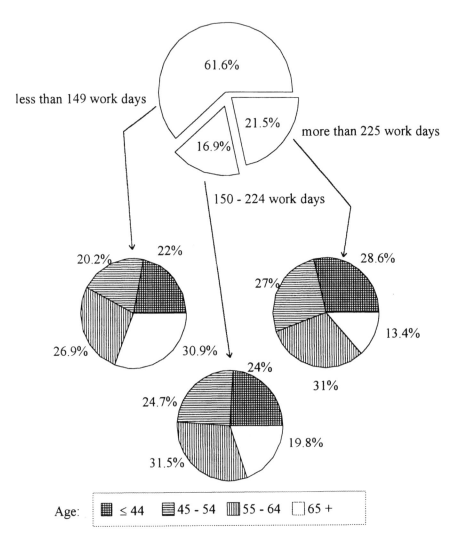

Figure 1.4: **Comparative structure of employment in agriculture and other sectors of the economy in Greece and the EU**

Source: *Farm structure*, 1993, Luxembourg, Eurostat

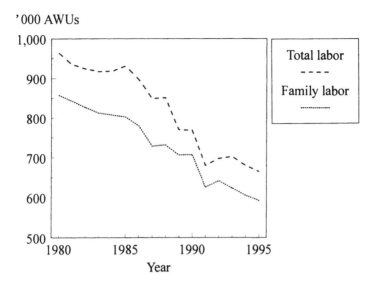

Figure 1.5: **The evolution of total and family labor in Greek agriculture, 1980-95**

Source: *Agricultural income 1995*, Luxembourg, Eurostat

Map 1.1 shows the location of the NUTSII regions, their common names in English as well as the regional names used by Eurostat in parentheses.

The absolute number of family labor including household heads, spouses and family members working on the farm differs among regions. The absolute number indicates the size of the region and highlights the relative importance of each region for Greek agriculture in terms of employment. The regions of Central Macedonia and Peloponnese are the largest regions while Attica (including Greater Athens), the Ionian and Southern Aegean Islands are the smallest regions. Participation of spouses and other family members varies from region to region according to the structure of the family farm, the nature of agricultural activities and alternative employment opportunities outside agriculture.

Map 1.2 shows the regional differences in the age structure of the holder. The Northern Aegean and Ionian Islands, Epirus, Attica and Peloponnese are top heavy with the elderly while regions in Macedonia and Thrace, the Southern Aegean Islands have a larger share of young farmers.

Map 1.3 shows the regional differences in the distribution of holdings according to the time devoted by the holder on the farm. The percentage of farmers working for less than 50% of their time on the farm is very high in Attica and all islands due to employment in tourism or other sectors of the regional economy. Regions in Macedonia, Thessaly and Epirus show the highest share of full-time farmers that should be attributed both to the profitability of agriculture in these regions as well as the lack of alternative employment opportunities.

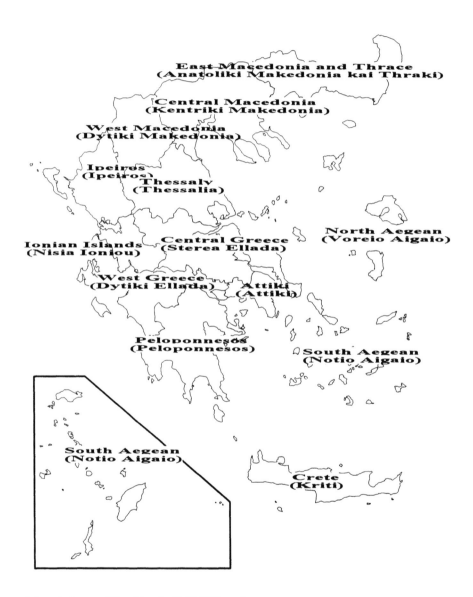

Map 1.1:   The Greek NUTSII regions

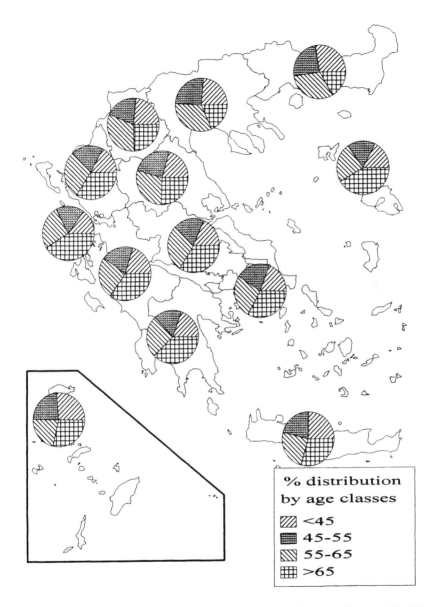

% distribution
by age classes

▨	<45
▦	45-55
▨	55-65
⊞	>65

Map 1.2: **Regional distribution of holdings by age classes of holder**
Source: *Farm structure*, 1993, Luxembourg, Eurostat

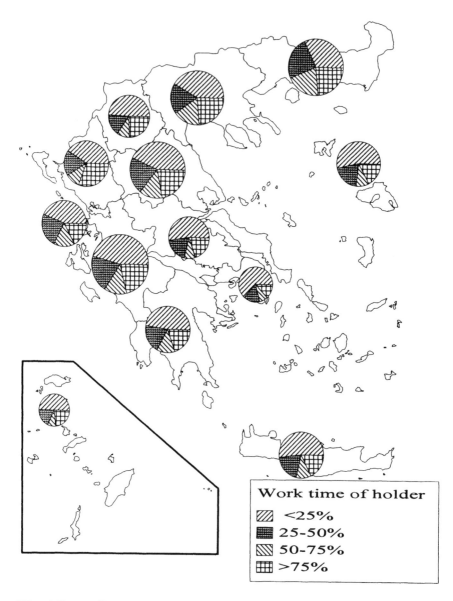

Map 1.3:   **Percentage allocation of work time of holder**
Source: *Farm structure*, 1993, Luxembourg, Eurostat

Thus it may be concluded that agriculture is still a major provider of employment in Greece that maintains the highest share of population active in agriculture among all EU member countries. An examination of demographic and human capital characteristics of the economically active population in Greek agriculture reveals certain problems related to a relatively aged structure, low skills and pluriactivity. The latter issue should be examined in relation to rural development policies aiming to create jobs in the non-agricultural sectors of the rural economy. The viability of rural households greatly depends on gainful off-farm activities by either the household's head or other family members.

Agricultural income

Crop output accounts for almost 70% of the country's final agricultural production. Almost 55% of the final agricultural production is comprised by six general categories of products, namely olive oil (13.7%), fresh vegetables (13.1%), cotton (7.6%), fresh fruits (7.0%), cereals (6.9%) and tobacco (6.1%), while milk, sheep and goat meat account for 8.5% and 6.7% respectively. The structure of final agricultural production is significantly different from the structure of the European Union as a whole and from other Mediterranean countries, revealing the importance of certain product market organization within the Community for Greece. The concept of agricultural income may be approached through different indices derived from national accounts. Eurostat, calculates three indicators based on the Economic Accounts for Agriculture (Eurostat, 1989). First, the net value added at factor cost in agriculture is calculated by taking the value of final agricultural output and deducting intermediate consumption, depreciation and taxes linked to production, and then adding subsidies. The second is the net income from the agricultural activity of total labor input that is deduced by subtracting rents and interest payments from net value added at factor cost. The third is the net income from the agricultural activity of family labor input and is calculated by deducting the compensation of employees from the net income of the agricultural activity of total labor input. Following these definitions, the latest available economic account for Greek agriculture is shown on table 1.4.

In the long term (1981-95), agricultural income in Greece has risen at an average rate of 1.7% per year, similar to that of the European Union as a whole. Over an even longer period (1973-95), two well defined sub-periods may be identified. First, the period up to 1980, before the country's accession to the EU, is characterized by strong and steady growth in incomes due to the rising prices of most crops with an aim to reach the corresponding European prices in view of the forthcoming accession, together with a marked rise in the volume of final output driven by the rapid expansion in the output volume of fiber plants (cotton in particular). These gains were maintained until the late 1980s, when a period of greater fluctuation ensued, which peaked in 1989 and 1991. Over the entire 1973-95 period, intermediate consumption grew fast which is still low compared to a European Union average of about 46%, although it accounts for about 27% of the final output value. The level of depreciation remained unchanged in the reference period accounting for almost 6% of the gross value added at market prices, compared to the European Union one that accounted for almost 29%. Subsidies rose by an average 9.6% per year in real terms while taxes linked to production rose by 3.3%. Net income of total labor and net income of family labor did not increase over the reference period due to the strong rise in real interest payments and the moderate fall in rental payments and the compensation of employees.

Table 1.4
Economic account for Greek agriculture, 1994

		values in million ECUs
	6,102	final crop output
+	2,620	final animal output
=	8,722	final output
-	2,286	intermediate consumption
=	6,436	gross value added at market prices
+	1,437	subsidies
-	253	taxes linked to production
=	7,620	gross value added at factor cost
-	380	depreciation
=	7,240	net value added at factor cost
-	258	rent and payments in cash or kind
-	451	interest
=	6,530	net income from agricultural activity of total labor input
-	433	compensation of employees
=	6,097	net income from agricultural activity of family labor input

Source: *Agricultural income 1995*, Luxembourg, Eurostat

Figure 1.6 presents the evolution of net value added at factor cost deflated by the implicit price index of gross domestic product at market prices and the same index with the net value added at factor cost divided by the volume of the total labor input measured in AWUs over the period 1973-95. Figure 1.6 also shows net income from the agricultural activity of total labor input and of family labor input deflated by the implicit price index of gross domestic product at market prices. Net income from agricultural activity is divided by the volume of total labor input in agriculture and by the volume of family labor input.

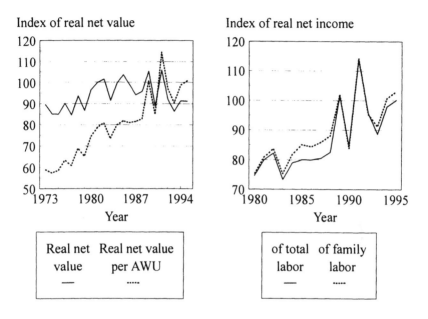

Figure 1.6: **The evolution of real net value and income in Greek agriculture**

Source: *Agricultural income 1995*, Luxembourg, Eurostat

In the European Union as a whole, the rise in agricultural income slowed down after 1992 due to the Common Agricultural Policy (CAP) reform. The reform aimed to adapt agricultural output to internal and external demand to achieve a better balance in the markets and a better competitive position for European Union agriculture. The measures focused on three main points: first, a reduction in the prices of agricultural products, the abolition of guaranteed prices for certain products and the reduction in the intervention prices for certain livestock; second, measures to control output such as the land set-aside scheme; and third, the granting of direct compensatory aid and the upgrading of certain types of existing aid to producers. The reform of the CAP was introduced in the Spring of 1992 and came into effect in the 1993-94 marketing year. The new orientation of the CAP has affected the structure and development of agricultural accounts. The fall in prices and volumes of output

27

are reflected by a decline in the value of final output and gross value added at market prices. According to the latest Farm Accountancy Data Network (FADN) results, the Greek average farm net value added per holding and family farm income per unit of unpaid labor are the second lowest among European Union member countries and almost 5 times lower than the highest ones (table 1.5). Family farm income per unit of unpaid labor also differs among the various types of farming with cereals ranking at the bottom and horticulture at the top. Mixed crop and livestock farms that account for the largest part of farms in Greece show the second lowest income (table 1.6).

Table 1.5
Farm net value-added and family income in Greece and other European Union countries, 1993-94

Average results per holding in '000 ECUs

	Farm net value-added	Farm net value-added per AWU	Family farm income per unit unpaid labor
Belgium	47.0	27.6	34.2
Denmark	37.6	27.0	6.0
Germany	27.1	16.7	15.7
Spain	16.8	13.4	14.4
France	33.9	19.8	20.3
Ireland	16.6	12.5	13.3
Italy	21.2	12.0	18.4
Luxembourg	35.9	22.1	31.4
The Netherlands	61.7	28.2	26.5
Portugal	2.6	1.8	1.4
United Kingdom	81.1	28.1	46.6
Greece	12.8	6.8	10.9

Source: *The agricultural situation in the Community 1995 report*, Luxembourg, Commission of the European Communities

Regarding employment in Greek agriculture it became evident that a large number of farm households depend on off farm employment in order to gain a viable income. An income measure which aims to be a proxy for the standard of living of the agricultural community will need to take into account income from other sources and not just that from farming activities. Eurostat's Total Income of Agricultural Households (TIAH) 1995 report (Eurostat, 1995) attempted an estimation of the composition of total income in 1980 and 1990.

The average real income from other forms of self-employment has been on a strong upward trend, doubling over the period. This is supported by empirical research showing that in non-pluriactive households income from agricultural activity accounts for almost 70% of total income while for pluriactive households this figure drops to 20% (Damianos, 1992). Wages in real terms have also been increasing but the strongest rise occurred with the income from property that experienced a four-fold increase (figure 1.7). Income distribution among agricultural households in Greece is less unequal than the corresponding one among non-agricultural households as revealed by studies using the Gini coefficient as a measure of disparity (Lazarides et al., 1989).

Table 1.6
Farm net value-added and family income by type of farming in Greece, 1993-94

Type of farming	Farm net value-added*	Farm net value-added per AWU*	Family farm income per unit unpaid labor*
Arable cultivation	13.1	7.1	10.3
Horticulture	14.6	8.1	13.9
Vineyards	10.7	6.1	9.6
Permanent crops	10.7	5.7	9.5
Dairy	20.3	9.3	17.5
Dry stock	15.6	8.2	14.9
Granivores	16.6	7.7	15.2
Mixed cropping	13.5	7.2	12.5
All types	12.8	6.8	10.9

* = average results per holding in '000 ECUs

Source: *The agricultural situation in the Community 1995 report*, Luxembourg, Commission of the European Communities

A very interesting comparison is that of income in the farm and non-farm sectors in Greece. The ratio of the per capita income in the two sectors, known as disparity ratio or relative income, increased in Greece for the period 1961-81 from 0.3 to almost 0.6, revealing a narrowing in the income gap between the two sectors (Caraveli-Ioannidis, 1987). In this period, the per capita income and labor productivity in Greek agriculture have been increasing faster than the corresponding magnitudes in the rest of the economy. In the post accession period the gap continued to narrow, with the influence of the developments in the terms of trade being more evident than that of agricultural labor productivity as related to that in the non-agricultural sector.

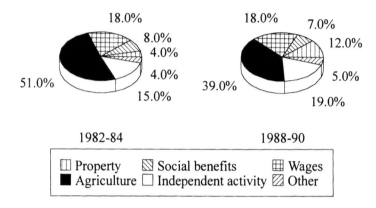

Figure 1.7: **Composition of the total income of households in Greece**
Source: *Total income of agricultural households*, 1995, Luxembourg, Eurostat

The agricultural share in the country's total domestic production is declining. The primary sector, including agriculture, fisheries and forestry accounted for more than 38% of the country's total production in 1951 and for less than 12% in 1993. Table 1.7 shows the gross domestic product in the different sectors of the economy and the gross domestic product per person employed. In the period 1970-93, gross domestic product in agriculture, in 1970 prices, grew by 29% while in the rest of the economy the respective growth was more than 110%. On the other hand, gross domestic product per person employed grew by a spectacular 145% in agriculture and almost 35% in the rest of the economy. This is attributed to the extreme exit rates observed in Greek agriculture, especially in the period 1970-80 that continued with lower rates up to the late 1980s and the early 1990s. It is very important to note that gross domestic product growth in agriculture has two distinct sub-periods of change, one up to 1980 that is the growth period and the other from 1980 onwards, which is a stagnation period. In this period the volume cf production remained stable while prices showed a very small increase up to 1988 and a decrease afterwards.

The share of the gross domestic product in agriculture and the other sectors of the economy is regionally differentiated (map 1.4). The Northern Aegean Islands, Attica and Western Macedonia show the lowest share while the regions of Thessaly, Crete and Peloponnese show the highest ones. This spatial differentiation may be attributed to various reasons including the development of the non-agricultural sectors, especially tourism in the island regions and manufacturing in the regions dominated by large urban centers (Attica, Central Macedonia, etc.). Another reason may be the share of the different agricultural products in the final regional agricultural production. Regions specializing in high-value production such as cash-crops, fruits or early vegetables under cover achieve high economic results (Thessaly, Peloponnese and Crete respectively). Some indications concerning the regional differentiation of the value of agricultural production per AWU show that all Greek regions are below the European average with the region of Epirus and the Ionian Islands ranking at the bottom of the spectrum and the region of Thessaly and Crete at the top (Commission, 1995).

Table 1.7
Gross domestic product in agriculture and the rest of the economy

Year	Agriculture		All other sectors	
	GDP*	GDP per capita	GDP	GDP per capita
1970	44,555	33,944	210,942	109,729
1975	54,229	43,733	283,100	133,601
1980	58,029	52,373	357,011	146,737
1985	58,212	57,636	388,887	156,809
1990	51,156	66,127	437,140	148,410
1993	57,493	83,239	448,768	148,132
% change				
1970-80	30.2	54.3	69.2	33.7
1980-93	-0.9	58.9	25.7	0.9
1970-93	29.0	145.2	112.7	34.9

* = GDP in million drachmas and GDP per capita in drachmas at 1970 prices

Source: *Statistical Yearbook of Greece*, 1970-93, Athens, NSSG

Agricultural trade

Greece presents a high degree of self sufficiency for most crop products and certain livestock products. This allows for extensive exports of agricultural products in the European Community and trade with third countries. Since the country's accession to the European Community, the degree of self-sufficiency in certain agricultural products has changed as a result of the changed agricultural prices.

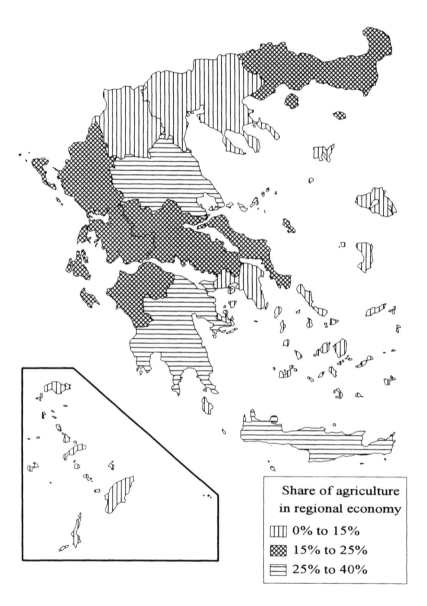

Map 1.4: **Percentage contribution of agriculture to regional output**
Source: *The agricultural situation in the Community 1994 report*, 1995,
Luxembourg, Commission of the European Community

Useful conclusions may be drawn by examining table 1.8 that presents the degree of self-sufficiency at accession and most recently in 1993/94. From the crop products, cereals, potatoes, fresh fruits and wine show a decline in self sufficiency, while rice, sugar and fresh vegetables show an increase. From the livestock products, fresh milk and eggs remained practically the same while all other products declined sharply. This should be attributed to the reduction in production resulted from cheaper Community imports of livestock products. The degree of self-sufficiency in the products referred to in table 1.8 as well as in other major crop products including olive oil, cotton and tobacco for which the country also presents a high degree of self-sufficiency allowed agricultural trade to develop as a major component of the country's trade in general.

Table 1.8
Self-sufficiency in certain agricultural products

Products	1979/80		1985/86		1993/94	
	Greece	EU10	Greece	EU12	Greece	EU12
Cereals	102	101	106	114	93	126[c]
Rice	115	83	140	79	171	75[a]
Potatoes	104	101	105	102	87	101
Sugar	92	124	96	129	101	135[c]
Fresh vegetables	118	98	150	107	158[b]	106[b]
Fresh fruit	163	83	136	87	122[b]	85[b]
Citrus fruit	137	44	163	75	145[b]	70[b]
Wine	105	105	116	105	87	97
Fresh milk	99	101	98	102	99	101
Cheese	92	105	88	106	81	106
Butter	70	119	49	110	38	104
Eggs	100	102	97	102	97	102
Beef and veal	50	98	32	107	31	107
Pig meat	87	100	69	102	65	106
Poultry meat	101	105	97	104	92	108
Sheep and goat meat	94	71	87	80	86	87

a =1990 or 1989/90
b =1988 or 1987/88
c =1992/93

Source: *The agricultural situation in the Community*, 1979-94, Luxembourg, Commission of the European Community

33

Agriculture has traditionally been the major exporter in the Greek economy. Agricultural exports accounted for more than 80% of the country's total exports in the period up to 1960 and for 19% of the country's agricultural gross product. In 1960, tobacco, currants and cotton alone out of all agricultural products accounted for 80% of all exports. Since the early 1960s however, other economic sectors took up exporting activities and the role of agriculture was continuously decreasing to account for a minimum of only 25% of total exports in 1982. It increased again to account for 32% of total exports and for more than 30% of the gross domestic agricultural product in 1994 (figure 1.8).

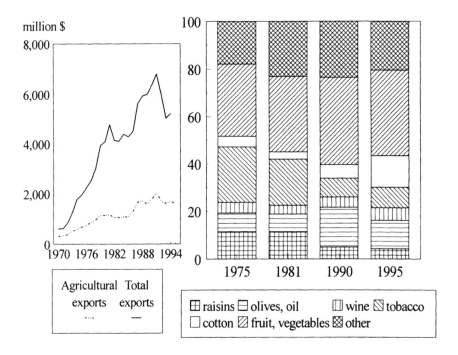

Figure 1.8: **Exports of agricultural products**
Source: *Monthly statistical bulletins*, Athens, Bank of Greece

Decreased exports are due to the decreased exports in traditional Greek products such as tobacco, currants and cotton that was only partly offset by increased exports in olives and olive oil and the spectacular growth in exports of fresh and preserved fruits and vegetables. The declining exports of tobacco, currants and cotton, however do not follow the same time pattern. Tobacco and currants increased their value of production up until 1981;this value was then halved by 1995. On the other hand, cotton shows decreasing exports up to the mid 1980s when its exports started rising to become in 1995 six times higher than in 1981. Other traditional Greek products such as wine maintained the same share of exports. Furthermore, agricultural exports are concentrated on products with unstable prices and relatively small demand elasticities. The changing structure of agricultural product exports is shown in figure 1.8.

Agricultural imports accounted for almost 15% of the country's total imports in 1995. Imports of agricultural products show a spectacular increase in absolute value but their share in the country's total imports is relatively stable at around 14%, with the exception of 1973, when international prices were very high, and reached 20% (figure 1.9).

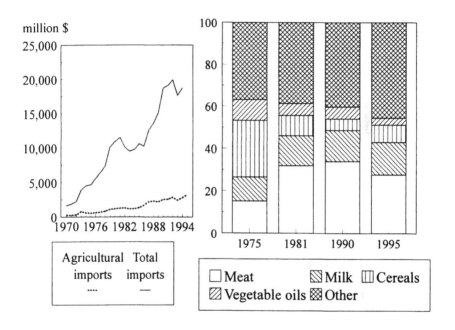

Figure 1.9:   Imports of agricultural products
Source: *Monthly statistical bulletins*, Athens, Bank of Greece

Agricultural product imports accounted for 15% of the country's gross domestic agricultural product in 1961 and for more than 40% in 1994. The increasing value of agricultural imports is due to the increase in imports of meat and live animals that accounted for almost 34% of imports of all agricultural products in 1990 and went down to 27.5% in 1995. The value of imports in milk and dairy products has also increased but its share in agricultural imports has not grown very much, from 11% in 1975 to 15% in

1995 (figure 1.9). The value of imported cereals and animal feed has tripled in the period 1975-95 but their share in agricultural imports has decreased from 27% in 1975 to 8% in 1995. The same pattern emerges for imported vegetable oil. In general, imports are concentrated on products with stable prices and high demand elasticities.

The trade balance for agricultural products has been positive up until 1979, with the exception of 1973. Since 1980, the balance of trade for agricultural products is increasingly negative (figure 1.10). In 1995, the trade balance for agricultural products was 1.6 billion U.S. dollars.

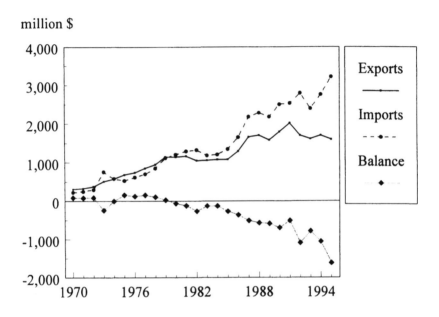

million $

Figure 1.10: **Imports, exports and trade balance for agricultural products**

Source: *Monthly statistical bulletins*, Athens, Bank of Greece

This evolution of the trade balance for agricultural products is unfavorable to Greece. First is the decreasing international demand for traditional Greek products such as oriental tobacco and currants due to changing consumption patterns. For example, smoking habits in the major tobacco importing countries and increasing international competition in the form of low priced products, as is the case of Turkish oriental tobacco, have reduced the demand for Greek oriental tobacco varieties (Dimara and Skuras, 1997). Second, are the changing consumption patterns in Greece as a result of increasing incomes and the consequent demand for meat and dairy products and other imported goods. A third reason concerns the low competitive position of Greek agriculture due to the high cost of production that is reflected in the pattern of total factor productivity (Fousekis, 1997). Lagging total factor productivity had serious implications on the country's trade advantage and hinders deep structural problems. Finally, is the issue of the structure of exports referring to the whole organizational framework of processing, packaging, transportation, and marketing infrastructure where Greek efforts have not been very successful.

Prior to the country's accession to the European Community in 1981, almost 50% of Greek agricultural exports and 27% of imports were directed to and came from European Community member countries while, 10 years later in 1990, the same figures had changed to 70% and 69% respectively. Since the early 1990s, Greek trade of agricultural products with Central and East European countries have grown in terms of both exports and imports. It is expected that the markets of these countries as well as markets of countries in the Balkans will import continuously increasing quantities of Greek agricultural products.

Investments in agriculture

Investments in agriculture are a good indicator of capital accumulation in the sector. Gross investments in agriculture (including forestry and fisheries) reached a peak in 1975 and then decreased continuously, with the notable exception of the 1980-85 period, to reach a lower level in 1993 than the corresponding 1960 one (table 1.9).

On the contrary, gross investments in the non-agricultural sector increased in the period 1960-80 and after a short fall following the country's accession to the EC, continued to rise up to 1993. The long term (1960-93) change of gross investments in the agricultural sector is negative while in the non-agricultural sector it is almost 260%. In the period following the country's accession to the EC, investments in agriculture increased while in the non-agricultural sector they decreased. This is due to the fact that some significant agricultural projects in irrigation and land reclamation were financed by Community funds and assistance was directed to certain private investments, especially in the field of processing and marketing of agricultural products. The share of agricultural investments decreased from 17% in 1960 to only 5% of total gross investments in 1993. At the same period, the ratio of gross investments to gross domestic product in agriculture decreased by a lower rate, from 17% in 1960 to 8% in 1993. Gross investment in agriculture and the rest of the economy is broken down according to its source into private and public investment and is presented in figure 1.11.

Table 1.9
Gross investments in agriculture and the rest of the economy

Year	Gross investments			GDP				
	1	2	3	4	5	6	7	8
	Agric.	Non-agr.	Total	Agric.	Non-agr.	1:3	1:4	2:5
1960	5,070*	24,051	29,121	29,863	99,338	0.17	0.17	0.24
1965	6,035	42,968	49,003	43,377	143,632	0.12	0.14	0.30
1970	7,523	63,140	70,663	47,058	210,942	0.11	0.16	0.30
1975	7,825	66,835	74,660	56,733	283,100	0.10	0.14	0.24
1980	6,169	86,536	92,705	60,499	357,011	0.07	0.10	0.24
1985	7,571	74,789	82,360	60,497	388,887	0.09	0.12	0.19
1993	4,872	86,420	91,292	60,223	446,038	0.05	0.08	0.19

* = all values are in million drachmas, at 1970 prices

Source: *National accounts*, 1960-93, Athens, NSSG

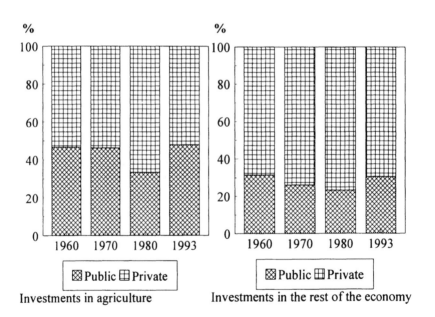

Investments in agriculture   Investments in the rest of the economy

Figure 1.11: The contribution of public and private investments in agriculture and the rest of the economy
Source: *National accounts*, 1960-93, Athens, NSSG

An examination of figure 1.11 reveals that the share of public sector investments in agriculture has decreased from 46.5% in 1960 to almost 33% in 1990, and increased to almost 48% in 1993. In the same time period, public investments in the rest of the economy are significantly lower than in the agricultural sector, accounting for almost 30% of the total investments in the non-agricultural sector in 1993 (figure 1.11).

If gross fixed capital formation in agriculture is broken down into type of investment, then it is obvious that the decrease is due to the reduced investments in buildings and construction and partly to the reduction in investments in machinery and transport equipment. The ratio of gross fixed capital formation to gross value added in agriculture reveals the rate of capital accumulation in the sector. This ratio decreased from 15.4% in 1975 to 13.6% in 1981, almost halved to 6.7% in 1991 and increased to 8.9% in 1993. In fact, the ratio of gross fixed capital formation to gross value added in agriculture at factor cost in Greece, is the lowest among all European Union countries.

Human capital investment in Greek agriculture is low not only because of the low level of general education among the Greek farming population but also due to the absence of technical training and efforts for further and continuing education in agriculture. Technical training and further education are mainly directed to the secondary sector of the economy and only a few state supported technical schools and the extension services of the Greek Ministry of Agriculture have attempted to provide technical training and further education in agriculture. Upon accession to the European Community, the country did not utilize directives concerning the provision of socio-economic guidance and the acquisition of occupational skills by persons engaged in agriculture and especially courses for training socio-economic counselors and the basic, further and advanced training of farmers. In the late 1980s however, Greece utilized European Union funds for providing training courses in rural Greece. These training schemes are either centrally designed by the Ministry of Agriculture, or locally designed by rural development groups. Courses are directed to young farmers and concerned with the acquisition of skills in agriculture and other activities that could be developed on farm and could diversify the rural economy, such as agritourism activities, further processing and alternative marketing of agricultural products, etc.

Agribusiness

Agriculture is highly related to other sectors of the economy in terms of input-output linkages that also influence employment and income. Agriculture is highly related to the agrifood industry and the agricultural input industries. The agrifood sector including the sectors of food, beverages and tobacco manufacturing is considered one of the most dynamic and promising sectors of the Greek economy, and in 1996 was the most profitable sector. According to the latest published Census of Manufacturing, the agrifood sector includes about 22,800 manufacturing establishments and employs 131,000 persons that accounts for almost 20% of the employment in the manufacturing sector. The food industry is the largest industry in the agrifood sector accounting for 80% of the employment followed by the beverages industry (10% of the sector's employment) and the tobacco manufacturing industry. The National Statistical Service of Greece computes an index of manufacturing production that is based on production data collected from about 1,800 factories selected by sector according to their size. The production value added of these factories during the base year of 1980 covered almost 77% of the total value added by the entire manufacturing industry and thus the index derived is a very reliable indicator of the performance of the different branches of economic activities and the manufacturing sector as a whole. The evolution of production indices for the manufacturing sector, and the industrial sectors of food, beverages and tobacco are shown in figure 1.12.

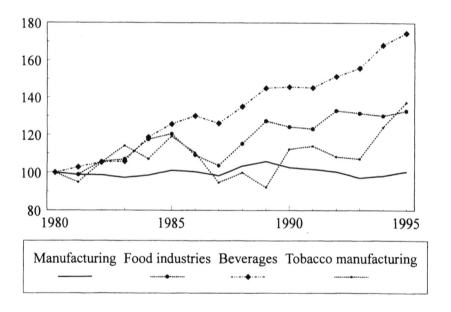

Figure 1.12: Evolution of manufacturing production indices, 1980-95
Source: *Statistical yearbook of Greece*, 1996, Athens, NSSG

Since 1980 the index for the whole manufacturing sector fluctuates around 100 while the indices for food, beverages and tobacco are continuously increasing except for 1986-87. The performance of the sectors is spectacular and the trend for the whole period is increasing.

Table 1.10 shows some financial data concerning large scale establishments (mean annual employment of more than 10 persons) in the food, beverages and tobacco sectors relative to large scale establishments in the whole manufacturing sector.

The three agrifood sectors account for 21% of the mean annual employment, 28.1% of the gross production value, 26% of value added and almost 30% of gross asset formation in the manufacturing sector. In 1996, from a sample of 281 medium and large size establishments, it was shown that the average return to own capital exceeded 15% while 120 of them presented profits of more than 100 million drachmas.

The importance of these figures is even more significant if the high multiplier effects of these sectors on the whole economy are taken into account. Multipliers show the impact caused by a change in a sector's product, employment and income by one, on the country's product, employment and income and indicates the input-output linkages of a sector with other sectors in the economy. The product, income and employment multipliers for agriculture and the food, beverages and tobacco sectors as estimated by Mattas (1992) are shown in table 1.11. The food sector has very high multipliers and consistently ranks among the 5 most important sectors of the Greek economy. The beverages and tobacco sectors, despite their relatively smaller product multipliers, show high income and employment multipliers indicating their importance to the Greek economy.

In the milk and dairy manufacturing sector there are 815 establishments of which 52% are small cheese factories processing from 1 to 3 tons of milk per day. There are only 10 large factories that treat more than 100 tons of milk per day and produce a range of milk and dairy products including pasteurized milk, cheese, yogurt and other specialty products. The sector's exports in 1994 were about 12,000 tons of cheese, 75% of which is "feta" cheese, and 5,000 tons of yogurt.

Eighty factories are active in the production of canned fruits and tomatoes, of which 10 produce both. Greece is the second major producer in the world of canned peaches, after the U.S and the largest exporter controlling almost 60% of world's exports. The 40 canned tomato producing factories have a capacity of more than 1,850,000 tons per year while the production for 1995 was 1,130,000 tons. More than 90% of the produce is exported, ranking Greece among the top five exporting countries. Six factories produce canned apricots and pears, and 40 factories produced marmalades and fruit pulp.

41

Table 1.10

Financial and other data concerning large-scale industries of the agrifood sector in Greece, 1981, 1992

Basic financial data	1981	1992	% of total manufacturing in 1992
Number of establishments			
food	583	1,146	13.7
beverages	112	186	2.2
tobacco	67	64	0.8
total manufacturing	3,961	8,346	100.0
Mean annual employment			
food	39,688	48,936	15.1
beverages	9,480	9,690	3.0
tobacco	8,282	9,569	2.9
total manufacturing	322,589	323,846	100.0
Gross production value[*]			
food	149,654	1,056,408	20.0
beverages	29,903	265,137	5.0
tobacco	27,261	165,546	3.1
total manufacturing	975,151	5,281,463	100.0
Value added			
food	35,645	334,783	17.0
beverages	11,146	118,760	6.0
tobacco	7,146	70,494	3.6
total manufacturing	284,854	1,973,655	100.0
Gross asset formation			
food	7,705	60,182	19.7
beverages	6,920	19,000	6.2
tobacco	878	10,058	3.3
total manufacturing	80,323	304,706	100.0

[*] = all values in million Greek drachmas

Source: *Statistical yearbook of Greece*, 1984, 1996, Athens, NSSG

Table 1.11

Multipliers for agriculture and the food, beverage and tobacco sectors

Sector	Product multiplier	rank*	Income multiplier	rank	Employment multiplier	rank
Agriculture	1.43	24	1.17	30	1.15	29
Food	2.22	8	7.03	3	7.35	3
Beverages	1.83	15	5.61	5	7.28	4
Tobacco	1.45	23	3.86	9	4.46	9

* = the rank of the respective figure among the 35 sectors of the economy

Source: Mattas, 1992

Olive oil is extracted in 3,270 small or medium establishments and more than 90% of production is consumed in the country and the rest is exported to European Union countries, especially Italy. There are also 92 factories producing 90,000 tons of canned olives, half of which are exported to Italy and countries in the Balkans. Wine production is another important activity with 303 active wine factories scattered all over the country. Exports of bottled wine are increasing and prospects are good taking into account that the stock of bottled wine in 1995 was the lowest in the last 15 years. On the contrary, production and exports of Greek packaged currants, sultan and other raisins are decreasing and the 12 large and 50 small currant factories face serious survival problems.

The tobacco and cigarette manufacturing sector is one of the most dynamic sectors of the Greek economy. The sector is characterized by intense exports, highly increasing profits, high gross margins of 28-34%, but a relatively low gross asset formation level. Investments since 1988 were mainly directed to machinery able to treat new tobacco varieties that were latter restricted (in 1992) and thus, a large part of the investments was never used or did not work at capacity levels. Tobacco manufacturing is a labor intensive activity taking place in rural areas at the tobacco producing prefectures and is thus very important to rural development. In the period 1991-94 under the national regional development grant assistance schemes, 11 large investment plans accounting for 2.3 billion Greek drachmas created 1,146 jobs, while in the period 1983-91, 41 investment plans with a value of almost 0.4 billion Greek drachmas created 132 jobs (Melissa, 1995).

From the agricultural inputs manufacturing sector, the most important branches are those of fertilizer and pesticide manufacturing and agricultural machinery. The Greek fertilizer industry had a peculiar structure as it was created and operated as a monopoly from the start of the century until the mid-

1960s. From the mid 1960s until 1992, the industry was comprised of four private and state owned factories that were obliged to sell their product to the state that marketed it through the Agricultural Bank of Greece and agricultural cooperatives. Prices until 1992 were set by the state, based on cost production estimates and were highly subsidized. The structure of the industry resulted in high production costs, lack of research on product improvement and development, and complete absence of marketing channels. Greek fertilizer industries now produce about 2.5 million tons of fertilizer, 90% of which is directed to the Greek market. It is very important to note that Greek industries produce very small quantities of organic fertilizers and do not produce liquid fertilizer due to their high cost. The 33 multinational and Greek pesticide industries in Greece do not produce chemical substances and are confined to the packaging and marketing of imported chemicals.

The Greek agricultural machinery industry produces a wide range of end products. The sector is decreasing and the trade balance for agricultural machinery has always been negative with some signs of recession since the early 1990s (Mergos and Psaltopoulos, 1996). The sector's prospects are not promising. Immediate restructuring along the lines of cost reduction, technological advancement, adoption of standards and improvement in marketing is needed.

Production patterns and consumption tendencies

Production of agricultural goods

As was shown earlier on in this chapter, the country's agricultural production consists mainly of crop products, while livestock output accounts for almost 30% of the value of total agricultural production. Despite the changes observed in food consumption and the demand for agricultural products, the composition of output remained unchanged (figure 1.13). Crop output composition has also remained unchanged consisting mainly of annual crops (60%) and perennial crops (40%). The spatial-temporal production pattern for most agricultural products in Greece, at the NUTSIII level, has been found to be very stable (Dimara, 1989). The volume of final output rose markedly during the 1970s with a rate that was not maintained during the 1980s. During the 1980s, the volume of final output increased by 0.7% per year on average, a little less than the European Union average. The real price for final output decreased at an accelerated rate, but the average fall (-2.2%) in the period 1980-95 is less steep than the average for the European Union as a whole, a fact that is, partly, attributed to the devaluation of the Greek drachma (figure 1.13).

million ECUs

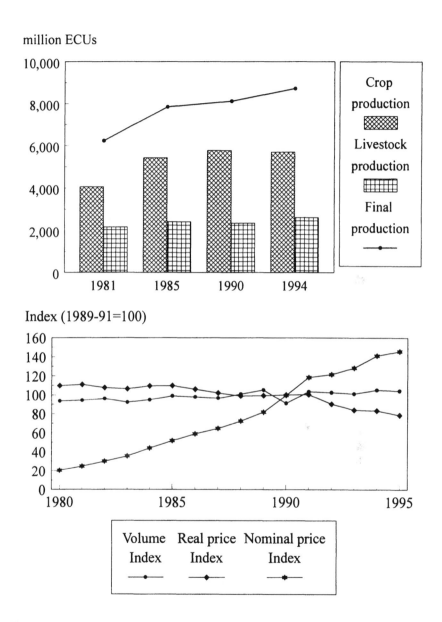

Figure 1.13: The evolution of agricultural production in Greece, 1980-95
Source: *Agricultural income 1995*, Luxembourg, Eurostat

The slump in real prices from 1985 onwards was brought about by the European Union's common market organization mechanisms that affected the various varieties of tobacco, cotton and olive oil (Dimara and Skuras, 1997). The impact of developments in the prices and volumes of cereals (a decline of -5.9% per year on average), potatoes (-0.6%), fresh fruit (-3.0%), sheep and goat meat (-4.2%) over the period 1981 to 1994, should not be ignored either.

The use of intermediate consumption in the period 1981-94 grew at a fast rate (1.6% annual average) and rose in 1995 to about 27% of the value of final output, compared to a European Union average of about 46%. Increases in the use of intermediate consumption are attributed mainly to the increases in the consumption of energy (an average of 4.5% per year), agrichemicals (6.4%) and feedstuffs (1.0%) (figure 1.14). On the other hand, the average price for intermediate consumption goods as a whole decreased by -1.7% per year on average, and therefore the real value of intermediate consumption remained relatively stable (figure 1.14).

Index (1989-91=100)

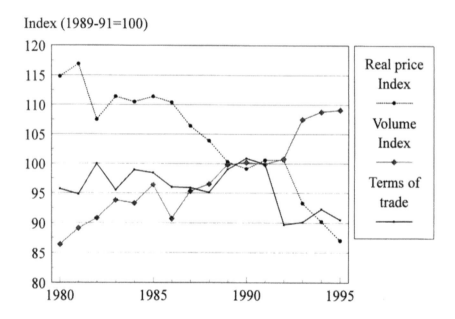

Figure 1.14: **Real value and volume indices of intermediate consumption and "terms of trade" of Greek agriculture**

Source: *Agricultural income 1995*, Luxembourg, Eurostat

The farmer's purchasing power may be examined by comparing the level of prices received for the final agricultural output to the prices paid by the farmer for intermediate consumption materials. Two measures that give, practically, the same results may be employed. The "cost-price squeeze" is calculated by dividing changes in the deflated index prices of the value of final agricultural production by changes in the deflated index prices of the value of inputs. The "terms of trade" which is calculated by dividing the index of nominal prices of final output by the index of nominal prices of intermediate consumption. The "terms of trade" deteriorated moderately over the period 1981-94 at an average of -0.8% per year (figure 1.14). An examination of the influence of high inflation on the terms of trade and prices received and paid, for Greek farmers in the period 1967-87 showed that pricing policies implemented by the Greek government had resulted in neutralizing the impact of inflation on the terms of trade (Daouli and Demoussis, 1989). The same research concluded that the observed variability of the terms of trade in this period should be attributed to real demand and supply factors.

From the set of other costs that are not included in the estimation of intermediate consumption, land prices and rents and agricultural wages are the most important. Farmland prices have exhibited substantial variability since the 1970s. Prices have risen sharply since 1970 and declined slowly and steadily in the 1980s. Current real agricultural land prices are almost two and a half times their 1970 levels. It has been found that, besides rental income, the non-constancy of the real interest rate played a significant role in land price determination and thus, farmland prices in Greece were not determined exclusively within the agricultural sector (Daouli and Demoussis, 1992). The interest rate policies of the Greek government in the inflationary 1970s had an impact on the farm sector and one of the side effects was the observed increase in farmland prices. A decomposition analysis of factor cost shares for Greek agriculture has confirmed the aforementioned remarks concerning land, capital and wages (Karagiannis et al., 1996).

Agricultural wages have increased substantially (tenfold) with an accelerated rate in the period 1975-85, and have tripled since then (Commission, 1996). Research findings indicate very high own-price elasticity (-0.38) for labor demand, positive relation to crop output prices (elasticity of 4.6), negative relation to livestock prices (elasticity of -2.08) a very limited relation to the prices of other inputs and a lack of substitution relation between capital and labor (Mergos, 1991). Labor shortages resulting from high out migration in the respective period caused labor supply uncertainty especially in peak seasons for most annual (cotton, tobacco) and perennial (fruits, grapes, olives) crops. Research undertaken in Greece (Damianos and Skuras, 1996b) has confirmed the existence of seasonal labor under supply, labor uncertainty in peak seasons does suggest the adoption of new labor saving technologies in the form of new crops rather than in the form of agricultural machinery or other capital investments.

The analysis undertaken by researchers regarding the supply response in Greek agriculture is very limited. This indicates the fact that policy decisions are not based on hard facts and thus choices are ad hoc and may be optimal only by chance. Table 1.12 summarizes most of the estimates on supply elasticities for agricultural products in Greece. The lack of research on the supply of perennial crop products is evident. The estimates presented vary significantly due to differences in the underlying economic model or estimation methods, data periods and econometric specification. Furthermore, cross-price elasticities of supply are rarely implemented due to the fact that supply response research is implemented in a single crop framework as opposed to a multi-product production framework ignoring the supply functions of other outputs and the demand functions for inputs (Mergos, 1991).

Table 1.12
Supply price elasticities for selective agricultural products

Product	Price elasticity		Source
	Short run	Long run	
Crop products			
Hard wheat	0.31	0.70	Baltas (1987)
Soft wheat	0.34-0.38	0.38-0.41	Baltas (1987)
Barley	1.33	2.80	Baltas (1987)
Oats	0.68	1.52	Pavlopoulos (1967)
Sugar beets	0.63	0.91	Apostolou and Varelas (1987)
Tobacco	0.35	0.96	Apostolou and Varelas (1987)
Cotton	0.65	3.31	Zanias (1981)
Cotton	0.70-0.85	1.04-1.22	Lianos and Rizopoulos (1988)
Peaches	0.17-0.26	0.39-1.18	Papadopoulou et al. (1992)
Livestock products			
Sheep/goat meat	0.16	----	Katranidis and Lianos (1992)
Milk	1.20	2.30	Pavlopoulos (1967)

Source: As indicated in the table

The main sources for the development of the Greek agricultural sector are the introduction of new and innovative agricultural technology and the increase of productivity. Technical advance may be measured by the rate of technical change from a production function, the rate of technical change from a value function and the rate of Total Factor Productivity (TFP). Mergos (1993) concluded that the evolution of TFP in Greek agriculture may be divided into two distinct periods, the one covering the years 1961 to 1976 and the other from 1976 to 1990. In the first period productivity increased by at least 3.5% per year, while in the second period productivity decreased equalizing increases in the first period. Fousekis and Papakonstantinou (1997) have extended the analysis to 1993 and found a considerable slowdown in productivity growth after 1981, the year of Greece's accession to the European Community. Productivity growth up to 1981 has been attributed to improvements in capital utilization and the resulting benefits in terms of reduction in the marginal production costs. Fousekis (1997) argues that the decrease in TFP is due to two reasons, first the increase in the rate of capital utilization in the early 1980s and 1990s which worked towards higher marginal production costs and lower productivity levels and second, the effort to restructure agriculture by directing producers to new crops and crop varieties which, in the short-run, decreases productivity due to higher adjustment costs. Mergos (1993) adds to the reasons that resulted in the decrease in TFP in the years after 1981. Since the mid 1970s, the Greek economy shows strong imbalance marked by internal and external deficits. Deficits are linked to high inflation rates, the devaluation of the Greek drachma and the decreased competitiveness of agriculture due to reduced investments. It is argued that the diffusion of innovations in agriculture is related to gross cumulative investments which may be one of the main reasons for the decreasing TFP. Furthermore, structural problems related to the land market may confront TFP increases. A great part of the farm households cultivate small parcels of land with low productivity and economic results. Finally, it is argued that the decreases in TFP after 1981 may be related to factors that did not allow the livestock sector to grow such as green currency parities and the negative monetary compensatory amounts. A recent survey about the main factors explaining the significant slowdown in productivity since the late 1970s, found that the market disequilibrium effect of capital is closely associated with a slowdown of investment in capital equipment over the same period (Mergos and Karagiannis, 1997).

Since the 1950s, Greece has gradually promoted intensification of agricultural production. Intensification is evident from an inspection of the evolution of yields for different crops (table 1.13). Increased yields are attributed to the use of inorganic fertilizers, pesticides and the mechanization of the production process. The total amount of nitrate fertilizer applied to Greek agriculture has increased five times since the early 1960s (Beopoulos and Skuras, 1997). The use of pesticides and other chemical substances has increased with the introduction of new high-yielding varieties of wheat, cotton and tobacco (Damianos et al., 1995). Furthermore, the number of tractors rose from 32,000 in 1961 to 227,000 in 1992 (table 1.13).

Agricultural mechanization in Greece is marked by the tremendous rate of adoption of virtually any type of electric water extraction pump, irrigation installations and especially drop and sprinkling units, and every type of tractors including harvesters threshers, mowers and sowing machines. As far as tractors are concerned, a recent survey revealed that increasing numbers of tractors are also associated with increased horsepower, despite the fragmented nature and the small size of the majority of farm enterprises (Mergos and Psaltopoulos, 1996).

It has been shown that technological developments in Greek agriculture are not Hicks-neutral, i.e., the change of the capital to labor ratio is not constant. In the period 1973-89 technological developments were found to aim at saving the non-reproducible factors of production i.e., labor and capital, and to intensify the use of machinery and chemical-biological inputs (Velentzas and Karagiannis, 1994). Moreover, the rate of technological changes in the use of machinery is higher than the corresponding one in the use of chemical-biological inputs, indicating that labor as a factor of production is more scarce than land. In general, the technological changes in the Greek agricultural sector, and the priority of technological innovation aim to save on land and labor and is in accordance with the Hayami-Ruttan model on technological changes in agriculture (Hayami and Ruttan, 1985).

There are great contrasts between various parts of the country concerning intensity of mechanization and adoption of modern technology. The number of tractors per hundred households is a good indicator of the regional variation of mechanization. Mechanization is high in regions where intensive agriculture is practiced, whereas it is limited in regions with extensive mountainous and less agriculturally developed regions (map 1.5).

Table 1.13
Yields and the use of tractors and chemical inputs in Greek agriculture

	1965	1975	1985	1995
Yield (Kg/ha)				
Wheat	1,687	2,377	2,142	2,859
Barley	1,660	2,321	1,867	2,503
Oats	1,249	1,618	1,503	1,905
Maize	1,728	3,834	8,612	10,135
Rice	4,405	5,085	6,157	8,052
Cotton	1,682	2,642	2,105	2,874
Sugar beets	39,454	60,405	60,617	65,090
Tomatoes	17,573	40,077	44,942	43,122
Oranges	14,816	18,469	18,466	28,913*
Apples	10,515	16,493	14,512	23,620*
Pears	14,092	23,769	20,000	24,690*
Peaches	12,260	12,045	15,998	21,809*
Number of:				
Tractors	49,093	152,889	183,410	228,051*
Electric pumps	20,719	52,305	97,047	128,083*
'000 tons of:				
Fertilizer (all)	828.1	1,383.9	2,239.7	2,000*
Pesticides (all)	134.2	399.9	------	------

* figures refer to 1992

Source: *Agricultural Statistics of Greece*, 1965-95, Athens, NSSG

Demand for agricultural products

Household expenditure on food has increased two and a half times since 1970 but the budget share of food decreased by almost half since the 1950s, as a consequence of increased incomes. Table 1.14 shows the budget shares for major food categories as well as the food share of the total household budget. The table clearly shows that there is a redistribution of food expenditures in favor of meat and fish. Changes in food consumption may be attributed to changes in prices, incomes and population. The prices for food products increased more rapidly than prices for durable and consumption goods and services such as health and personal care, transport and communication but less rapidly than prices for alcoholic drinks and tobacco, clothing and

51

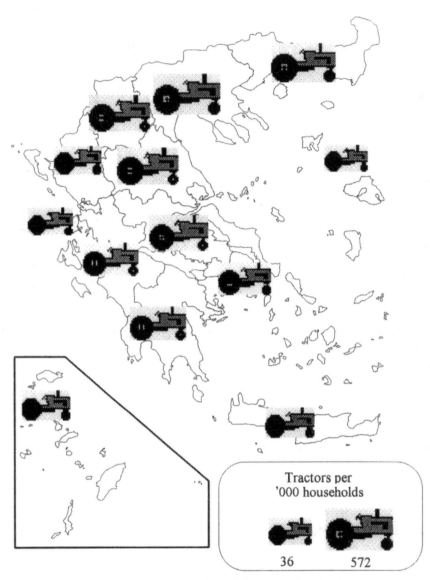

Tractors per '000 households

36   572

Map 1.5: **The regional dimension of farm mechanization**
Source: *Agricultural Statistics of Greece*, 1992, Athens, NSSG

footwear and housing (figure 1.15). Among food items, the prices of fruits and vegetables have greatly increased while the price of meat and dairy products have modestly increased. Income should be considered as a major factor influencing food consumption patterns in Greece. Per capita income in constant prices has increased almost three times over the period 1960-95, while the change in population probably cannot explain all the observed changes in food consumption. Changes in consumption patterns are important if they are materialized to correspond to shifts in production patterns.

As was shown earlier on in this chapter, this link has not occurred in Greek agriculture with adverse impacts on the trade balance of the sector as it concerns livestock and dairy products. Observed changes in food items' budget shares are divided into three components: the total substitution, the budget (total expenditure), and the habit formation effect. It has been found that there is no single force responsible for the evolution of consumption patterns in post-war Greece as the effects that mainly explain this evolution of the consumption patterns vary among food items (Karagiannis and Velentzas, 1997).

Table 1.14
Budget shares for main food categories in Greece

| Food category | Average budget shares | | | | |
	1950s	1960s	1970s	1980s	1990s
Bread and cereals	0.24	0.15	0.09	0.07	0.06
Meat	0.11	0.16	0.24	0.27	0.26
Fish	0.04	0.06	0.04	0.07	0.08
Milk, cheese and eggs	0.14	0.16	0.15	0.16	0.17
Oils and fats	0.10	0.09	0.08	0.06	0.05
Fruit and vegetables	0.27	0.26	0.29	0.28	0.26
Other food	0.10	0.11	0.10	0.10	0.11
Food share	0.45	0.41	0.31	0.27	0.25

Source: *National accounts of Greece*, 1950-1993, Athens, NSSG

Index (1980=100)

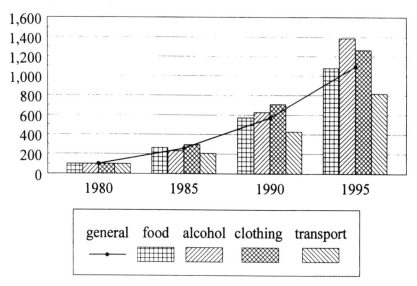

Figure 1.15: Consumer price indices for major expenditure categories
Source: *Statistical Yearbook of Greece*, 1980-95, Athens, NSSG

A few studies have attempted to examine consumer behavior and the demand for agricultural products in the Greek economy. Early work was based on the estimation of demand equations for individual products (Sakellis, 1982) or on the relation of demand and supply rather than the relation of demand and utility (Sakellis, 1982). An integrated study of the whole food sector as opposed to the non-food sector, using the Rotterdam model, revealed that the price elasticity of demand for all groups of food products is negative, the income elasticity of demand for food products is significantly lower than for non-food products, while the highest income elasticities were estimated for fish (1.77) and meat (1.07) and the lowest one for cereals and fats and oil (Demoussis, 1985). Latter studies on food demand in Greece using the Almost Ideal Demand System (AIDS) analysis confirmed the aforementioned conclusions (Andrikopoulos et al., 1987; Rigas, 1987; Mergos and Donatos, 1989a, 1989b; Demoussis and Drakos, 1995).

Demand for agricultural products as inputs, Mergos and Yotopoulos (1988) estimated derived demand of feed inputs in the Greek livestock sector by using a multiple-input multiple-output cost function and found that the responsiveness of farmers to input and output prices is very high (Mergos, 1991).

Bibliography

Data sources

Bank of Greece (1997), *Monthly Statistical Bulletin* (1960-1997), Economic Research Division, Bank of Greece: Athens.

Commission of the European Communities (1997), *The Agricultural Situation in the Community (1975-1996)*, Office for Official Publications of the European Communities: Luxembourg.

Eurostat (1996), *Agricultural Income*, Office for Official Publications of the European Communities: Luxembourg.

Eurostat (1995), *Total Income of Agricultural Households*, Theme 5, Series D, Office for Official Publications of the European Communities: Luxembourg.

Eurostat (1993), *Farm Structure: 1993 Survey, Main Results*, Theme 5, Series C, Office for Official Publications of the European Communities: Luxembourg.

Eurostat (1990), *Farm Structure: 1989 Survey, Main Results*, Theme 5, Series C, Office for Official Publications of the European Communities: Luxembourg.

Eurostat (1989), *Manual on Economic Accounts for Agriculture and Forestry*, Theme 5, Series E, Office for Official Publications of the European Communities: Luxembourg.

National Statistical Service of Greece (1997), *Agricultural Statistics of Greece (1960-1996)*, National Statistical Service of Greece: Athens.

National Statistical Service of Greece (1997), *Farm structure survey (1977-1993)*, National Statistical Service of Greece: Athens (available on electronic media).

National Statistical Service of Greece (1996), *Statistical Yearbook of Greece 1994/95*, National Statistical Service of Greece: Athens.

National Statistical Service of Greece (1995), *National accounts (1960-1993)*, National Statistical Service of Greece: Athens.

National Statistical Service of Greece (1995), *Labor Force Survey 1994*, National Statistical Service of Greece: Athens.

National Statistical Service of Greece (1991), *Census of Agriculture*, National Statistical Service of Greece: Athens (available on electronic media).

National Statistical Service of Greece (1991), *Census of Population*, National Statistical Service of Greece: Athens (available on electronic media).

European Union legislation - Policy documents

Commission of the European Communities (1996), *Structural Funds and Cohesion Fund, 1994-1999: Regulations and Commentary*, Office for Official Publications of the European Communities: Luxembourg.

Commission of the European Communities (1995), *Study on Alternative Strategies for the Development of Relations in the Field of Agriculture Between the EU and the Associated Countries with a View to Future Accession of These Countries*, CSE (95) 607, Office for Official Publications of the European Communities: Luxembourg.

References

Alivisatos, B. (1932), *La Reforme Agraire en Grece*, Recueil Sirey: Paris.

Anastasiades, H. (1911), 'The administration of the "National Estates" of the contemporary State', Athens. (in Greek)

Andrikopoulos, A., Brox, J. N. and Georgakopoulos, T. A. (1987), 'Short-Run Expenditure and Price Elasticities for Agricultural Commodities: The Case of Greece, 1951-1983', *European Review of Agricultural Economics*, Vol. 14, pp335-346.

Apostolou, N. and Varelas, E. (1987), *Production Functions for Selected Agricultural Crop Products*, Agricultural Bank of Greece: Athens. (in Greek)

Baltas, N. C. (1987), 'Supply Response for Greek Cereals', *European Review of Agricultural Economics*, Vol. 14, No. 2, pp. 403-10.

Beckinsale, M. and Beckinsale, R. (1975), *Southern Europe: The Mediterranean and Alpine Lands*, Hodder and Stoughton: London.

Beopoulos, N. and Skuras, D. (1997), 'Agriculture and the Greek Rural Environment', *Sociologia Ruralis*, Vol. 37, No. 2, pp. 255-69.

Caraveli-Ioannidis, H. (1987), 'Farm Income Disparity in Greece and Membership of the EC', *European Review of Agricultural Economics*, Vol. 14, pp. 239-49

Carter, W. (1968), 'Population Migration to Greater Athens', *Tijdschrift voor Economische en Sociale Geographie*, Vol. 59, pp. 100-105.

Damianos, D. (1992), 'Income Distribution and the Planning of Taxation(?) Policy for Farmers', in the 1st National Conference of Agricultural Economics Proceedings, *Greek Agriculture in the 90's: Economic and Social Perspectives*, Ministry of Agriculture: Athens.

Damianos, D., Demoussis, M. and Kasimis, C. (1991), 'The Empirical Dimension of Multiple Job-Holding Agriculture in Greece', *Sociologia Ruralis*, Vol. 31, No. 1, pp. 37-47.

Damianos, D., Dimara, E. and Skuras, D. (1995), 'Land Use Strategies Under Market Restructuring', in Albisu, L. M. and Romero, C. (eds), *Environmental and Land Use Issues: An Economic Perspective*, Wissenschaftsverlag Vauk Kiel KG: Kiel.

Damianos, D. and Skuras, D. (1996a), 'Farm Business and the Development of Alternative Farm Enterprises: an Empirical Analysis in Greece', *Journal of Rural Studies*, Vol. 12, No. 3, pp. 273-283.

Damianos, D. and Skuras, D. (1996b), 'Unconventional Adjustment Strategies for Rural Households in the Less Developed Areas in Greece', *Agricultural Economics*, Vol. 15, pp. 61-72.

Daouli, J. and Demoussis, M. (1995), 'Off-farm Work in Greek Agriculture: An Empirical Application', *Investigation Agraria - Economia*, Vol. 10, No. 1, pp. 77-89.

Daouli, J. and Demoussis, M. (1992), 'Rents, Interest Rates and Real Agricultural Land Prices: An Application to a Greek Province', *European Review of Agricultural Economics*, Vol. 1, pp. 417-25.

Daouli, J. and Demoussis, M. (1989), 'The Impact of Inflation on Prices Received and Paid by Greek Farmers', *Journal of Agricultural Economics*, Vol. 40, pp. 232-39.

Demoussis, M. (1985), 'Demand for Seven Food Commodities in Greece: An Application of the Rotterdam Model', *Review of Agricultural Studies*, Vol. 1, pp. 3-28. (in Greek)

Demoussis, M. and Drakos, P. (1995), 'Characteristics of Meat Demand in Greece', in Janssen, E. (ed.), *Advances in Stochastic Modeling and Data Analysis*, Kluwer Academic Publishers: Dordrecht.

Dicks, T. R. B., (1967), 'Greater Athens and the Greek Planning Problem', *Tijdschrift voor Economische en Sociale Geographie*, Vol. 58, pp. 271-75.

Dimara, E. (1989), 'L' Agriculture Greque: Etude Chronologique et Régionale de la Repartition des Cultures de 1970 a 1981', *Les Cahiers de l'Analyse des Données*, Vol. 14, No. 2, pp. 211-38.

Dimara, E. and Skuras, D. (1997), *Tobacco Growth in Greece*, Omvros: Athens. (in Greek)

Efstratoglou-Todoulou, S. (1990), 'Pluriactivity in Different Socio-economic Contexts: A Test of the Push-Pull Hypothesis in Greek Farming', *Journal of Rural Studies*, Vol. 6, No. 4, pp. 407-13.

Fousekis, P. (1997), 'Internal and External Scale Effects in Productivity Analysis: A Dynamic Dual Approach', *Journal of Agricultural Economics*, Vol. 48, No. 2, pp. 151-66.

Fousekis, P. and Papakonstantinou, A. (1997), 'Economic Capacity Utilization and Productivity Growth in Greek Agriculture', *Journal of Agricultural Economics*, Vol. 48, No. 1, pp. 38-51.

Hayami, Y. and Ruttan, V. W. (1985), *Agricultural Development: An International Perspective*, Johns Hopkins University: Baltimore.

Karagiannis, G. and Velentzas, K. (1997), 'Explaining Food Consumption Patterns in Greece', *Journal of Agricultural Economics*, Vol. 48, No. 1, pp. 83-92.

Karagiannis, G., Katranidis, S. and Veletzas, K. (1996), 'Decomposition Analysis of Factor Cost Shares: The Case of Greek Agriculture', *Journal of Agricultural and Applied Economics*, Vol. 28, pp. 369-79.

Kasimis, C. and Papadopoulos, A. (1997), 'Family Farming and Capitalist Development in Greek Agriculture', *Sociologia Ruralis*, Vol. 37, No. 2, pp. 209-27.

Kasimis, C. and Papadopoulos, A. (1994), 'The Heterogeneity of Greek Family Farming: Emerging Policy Principles', *Sociologia Ruralis*, Vol. 34, No. 3/4, pp. 206-28.

Katranidis, S. and Lianos, T. (1992), 'Sheep and Goat Meat Market in Greece and Impacts from the CAP Reform', in the 2nd National Conference of Agricultural Economics Proceedings, *Greek Agriculture in the Framework of the European Common Market*, Department of Agricultural Economics of the University of Thessaloniki: Thessaloniki. (in Greek)

Lazarides, P., Papailia, T. and Sakeli, M. (1989), *Income Distribution in Greece*, Studies No. 35, Agricultural Bank of Greece: Athens. (in Greek)

Lianos, T and Rizopoulos, G. (1988), 'Estimation of Social Welfare Weights in Agricultural Policy: The Case of Greek Cotton', *Journal of Agricultural Economics*, Vol 39, No. 1, pp. 61-8.

Mattas, K. (1992), 'Macro-economic decisions and the agricultural sector', in the 2nd National Conference of Agricultural Economics Proceedings, *Greek Agriculture in the Framework of the European Common Market*, Department of Agricultural Economics of the University of Thessaloniki: Thessaloniki. (in Greek)

Melas, G. and Delis, D. (1981), *Agricultural Wages and Employment in Agriculture*, Studies No. 11, Agricultural Bank of Greece: Athens.

Melissa, Th. (1995), 'Production and trade of tobacco', *Foundation of Economic and Industrial Research,* Athens. (in Greek)

Mergos, G. (1993), 'Total Factor Productivity in Agriculture: The Case of Greece 1961-1990', Paper Presented at the *VIIth European Agricultural Economics Association Congress*, Stresa, 6-10 September.

Mergos, G. (1991), *Output Supply and Input Demand in Greek Agriculture: A Multi-Output Profit Function Approach*, Center of Planning and Economic Research: Athens.

Mergos, G. and Donatos, G. (1989a), 'Demand for Food in Greece: An Almost Ideal Demand System Analysis', *Journal of Agricultural Economics*, Vol. 40, No. 3, pp. 178-84.

Mergos, G. and Donatos, G. (1989b), 'Consumer Behavior in Greece: An Application of the Almost Ideal Demand System', *Applied Economics*, Vol. 21, No. 7, pp. 983-99.

Mergos, G. and Karagiannis, G. (1997), 'Sources of Productivity Change Under Temporary Equilibrium and Application to Greek Agriculture', *Journal of Agricultural Economics*, Vol. 48, No 3, pp. 313-29.

Mergos, G. and Psaltopoulos, D. (1996), *The Agricultural Machinery Industry and Agricultural Mechanization*, Foundation of Economic and Industrial Studies: Athens.

Mergos, G. and Yotopoulos, P. (1988), 'Demand for Feed Inputs in the Greek Livestock Sector', *European Review of Agricultural Economics*, Vol. 15, No 1, pp. 1-17.

Moisidis, A. (1985), *Agricultural Society in Contemporary Greece*, Institute for Mediterranean Studies: Athens. (in Greek)

Papadopoulou, E. and Psaroudas, S. (1992), 'Specification and Estimation of Peach Supply Econometric Models with Co-integration Pre-specification in the 2nd National Conference of Agricultural Economics Proceedings, *Greek Agriculture in the Framework of the European Common Market*, Department of Agricultural Economics of the University of Thessaloniki: Thessaloniki. (in Greek)

Pavlopoulos, P. (1967), *Supply Functions of Agricultural Products: A Quantitative Investigation*, Center of Planning and Economic Research: Athens. (in Greek)

Rigas, K. (1987), 'Food Consumption in Greece: An Application of the Almost Ideal demand System', *Review of Agricultural Studies*, Vol. 2, pp. 41-63. (in Greek)

Sakellis, M. (1982), 'Econometric Investigation of the Factors Influencing Demand for Basic Agricultural Products', *Review of Agricultural Studies*, pp. 2-17.

Simonide, B. (1923), 'La Question Agraire en Grèce', *Revue d' Economie Politique*, Vol. 37, pp. 767-809.

Stefanides, D. (1948), *Agricultural Policy*, Athens.

Thompson, K. (1963), *Farm Fragmentation in Greece*, Research Monograph Series No. 5, Center of Planning and Economic Research: Athens.

Velentzas, K. and Karagiannis, G. (1994), 'An Analysis of Technical Changes in Greek Agriculture During 1973-1989', in the 2nd National Conference of Agricultural Economics Proceedings, *Greek Agriculture in the Framework of the European Common Market*, Department of Agricultural Economics of the University of Thessaloniki: Thessaloniki. (in Greek)

Vernicos, N. (1973), *L' Evolution et les Structures de la Production Agricole en Grèce*, Dossier de Rechèrche, Université de Paris VIII: Paris.

Wagstaff, J. M. (1968), 'Rural Migration in Greece', *Geography*, Vol 53, pp. 175-79.

Zanias, G. (1981), 'An Estimation of the Supply Response of Cotton in Greece from 1950-1979', *Oxford Agrarian Studies*, Vol. 10, pp. 196-211.

2 Rural environment and agricultural institutions

Agriculture and the rural environment

Farm and rural typologies in Greece

The Farm Accountancy Data Network (FADN) of the European Union provides a classification of farm holdings according to their production structure. Such a classification facilitates the interpretation of Community surveys in a common framework and allows for the comparison among member countries. The FADN typology is based on two essential economic characteristics of the holdings in each member country, that of farm type and the economic size of the holding. The standard for measuring farm type and the economic size of the holding is the gross margin. The gross margin is the difference between the monetary value of agricultural production (gross production) and the main proportional specific costs corresponding to the production concerned. It is therefore, an indicator of the potential net holding income, having the advantage of not favoring input-intensive types of production. According to FADN typology the economic size of the holding is the aggregate standard gross margin (SGM) calculated for every hectare and animal belonging to a holding. The economic size is expressed in European size units (ESU) which corresponds to some fixed amount of thousand ECUs (for example 1,200 ECUs for 1992), and is adjusted regularly to take into account, in monetary terms, the overall agri-economic trends throughout the European Union. The type of farming of a holding is determined by the relative contribution of the various types of production practiced to its total standard gross margin.

Table 2.1 presents the percentage share for the main farm types according to FADN typology in Greece and the European Union, and the percentage change in the number of farms of each farm type in Greece. The largest farm type is that of permanent crops including holdings with vineyards, fruit and citrus fruit plantations, olive groves and various permanent crops combined. This farm type is the only one that presents a positive change in the number of holdings in the period 1981-93, due to the doubling of holdings with olive groves.

Table 2.1
Farm types in Greece and the European Union

Farm Types	Greece			EU12
	% share of each type		% 1981-93	% share of each
	1981	1993	change	type 1993
Specialist Farms				
Field crops	29.8	24.6	-31.6	20.2
Horticulture	2.7	1.8	-44.3	2.8
Permanent crops	37.1	48.3	7.8	32.7
Grazing livestock	5.0	5.7	-6.3	22.3
Granivores	0.8	0.3	-67.9	1.4
All specialist farms	74.5	80.4	-10.8	79.4
Mixed Farms				
Cropping	17.2	11.4	-45.1	9.6
Livestock	2.1	2.3	-10.4	3.6
Crops-livestock	5.3	5.6	-13.4	7.4
All mixed farms	25.5	19.6	-36.2	20.6
All farms	100.0	100.0	-17.2	100.0

Source: *Farm structure surveys*, 1987 and 1993, Luxembourg, Eurostat

Field crops account for almost one quarter of the holdings in Greece but their number decreased by almost 32% due to a 50% reduction in the number of specialist cereals holdings. Specialist grazing livestock farms increased in number up to 1987 and then greatly decreased but maintain the same share of around 5%. In the European Union as a whole, specialist grazing livestock farms account for more than 22% of the total number of holdings. Specialists in granivore production holding present the highest decrease of almost 68% due to the reduced number of farms raising pigs and poultry in the period

1981-93. The number of all specialist farms in Greece decreased by more than 10% but increased its share in the total number of holdings approaching the European Union average figures. Finally, mixed cropping farms decreased both in number and in terms of percentage share approaching the respective European Union figures. This is a clear indication of increased specialization in Greek agriculture as mixed cropping production systems gradually decrease and are spatially confined to mountainous, island and less favored areas where agriculture still follows a survival strategy serving the needs for food of the rural household and the local community. The FADN typology according to ESUs will be presented in the following section dealing with the farm structure in Greek agriculture.

Besides the FADN typology according to the type of farm, a few approaches to construct agricultural typologies have been attempted by Greek scientists. Recent research carried out on alternative farming systems in the less favored regions of the EC attempted, among others, to produce a typology of farm holdings in certain less favored areas of Greece (Damianos et al., 1994). The typology was based on FADN and survey data collected in a less favored prefecture of Greece and grouped farm holdings operating in a similar manner, and following similar paths of business development over the years prior to the study. The working hypothesis underlying this typology was that holdings grouped in the same cluster are likely to adjust and adapt in the external environment in a similar fashion. Thus the typology may be used as a tool for the evaluation of the adaptive capacity of holdings and the assessment of impacts of policy changes on the economic viability of farm holdings. The typology created four main clusters of farm holdings, based on economic, social, demographic and other characteristics of the farm holding, the household's head and the members of the family farm. The first cluster was named "Entrepreneur Farmers", comprised of young farmers with relatively large holdings, expanding business, medium intensity in the use of resources, good agricultural and general education and accounted for almost 5% of the farm holdings in the area. The second cluster named "Under Pressure Farmers", comprised of young and middle aged farmers, medium sized (10 ha) holdings, stable or marginally declining business, intensive use of resources, fair agricultural education and accounted for almost 20% of the holdings in the area. The third cluster named "At Risk Farmers", comprised of elderly farmers with small or medium holdings (5 ha), casual involvement in off-farm jobs, declining business, intensive use of resources, poor agricultural education and accounted for 41% of the area's total number of holdings. The fourth cluster named "Pluriactive Farmers" comprised of middle aged or elderly farmers with very small holdings (2 ha) extensive involvement in off-farm job, declining farm business, very intensive use of resources, poor agricultural education and accounted for 34% of the holdings in the area. This fourth group was further sub-divided into "Pluriactive Winding Down Farmers" accounting for 35% of the farms in the fourth group or almost 12% of the farms in the area, and the "Pluriactive Entrepreneur Farmers" accounting for 65% of the farms in the fourth group and almost 22% of the total number of farms.

A geographic typology of rural areas in Greece was recently constructed under a European research project evaluating the impact of public institutions in less favored rural and coastal regions of the Community (Dimara and Skuras, 1996). The aim of this typology was to reveal the spatial dimension of rural problems in Greece and to identify problematic rural areas that are in immediate need of assistance and should become the subject of integrated rural development programs. This typology attempted to classify Greek prefectures (NUTSIII areas), according to the socio-economic performance of their rural areas and was based on a range of demographic, economic, service provision and social well-being indicators. The typology produced two clusters including prefectures that are dominated by large urban centers and having a predominantly urban character, and four clusters including predominantly rural prefectures. The first cluster comprised of 21 rural dynamic prefectures that were characterized by a dynamic agricultural sector, relatively high standards of living and well being and low unemployment rates. The second cluster comprised of 9 rural dynamic prefectures that were characterized by increasing employment in the non-primary sectors of the rural economy. The third cluster was made up of 8 mountainous and island areas that exhibit a static or declining agricultural sector and have experienced high out-migration levels, show no sign of development in the other sectors of the economy and the standard of living is very low. The fourth cluster contained 6 island prefectures that despite the static or declining primary sector and the extreme desertion problems showed an increase in economic performance due to tourism activities and a potential for further development.

Farm structure

Farm structure is a major determinant of agricultural production as it places restrictions on the farmer's choice of farming systems, controls the availability of capital and credit for investment in new forms of production and the development of vertically integrated businesses. Greek agriculture is characterized by an unfavorable farm structure with numerous small and fragmented farm holdings. The mean farm size is very small (between 3.6 and 4.3 ha) and has not increased significantly over the last 30 years. In comparison to other European Union member countries, Greece has the lowest mean farm size which is almost one quarter of the European mean for the 12 member countries, before the last enlargement (figure 2.1). Earlier research in Greece has found that the average farm size varies in different groups of the agricultural population according to the socio-economic characteristics of the farming population. At a regional level it is positively related to industrialization and mechanization and negatively related to population pressure on agricultural resources (Lianos and Parliarou, 1986).

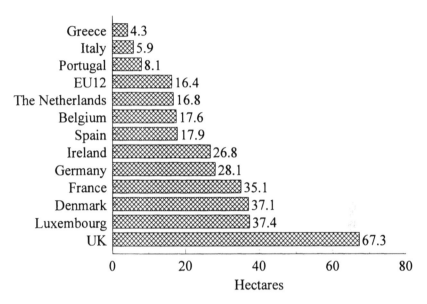

Figure 2.1: **Average farm size in various European Union member countries, 1993**

Source: *Farm structure survey*, 1993, Luxembourg, Eurostat

Table 2.2 shows the long term changes in the size distribution of number of holdings and the corresponding utilized agricultural area for Greece in 1961 and 1991. The proportion of very small (less than 1 ha) holdings increased, the proportion of medium sized farms (1 to 10 ha) remained relatively stable and the proportion of large farms (more than 10 ha) increased by more than 130% revealing a trend for the concentration of holdings to larger size groups and an increase of subsistence farms. The proportion of the utilized agricultural area by large holding (more than 10 ha) increased by almost 100%.

Table 2.3 compares the size distribution of number of holdings in Greece and the European Union of 12 as a whole, and also shows, within Greece, the differences in the size distribution in mountainous, less-favored and other areas. Holdings in Greece are concentrated in the less than 5 ha size groups, a situation that is more evident in the mountainous and less-favored areas of the country. It should be noted that Italy, Greece and Spain together account for more than 80% of the under 2 ha holdings in the European Union, Italy alone accounted for 50%, Spain for 16% and Greece for almost 14%. Less favored and mountainous areas present a significantly different size structure than other areas in Greece.

Table 2.2
Long-term evolution of the size distribution of number of holdings and utilized agricultural area

Size classes (ha)	% Number of holdings		% Utilized agricultural area	
	1961	1991	1961	1991
without land	1.4	1.0	0	0
<0.9	22.7	25.8	3.6	3.2
1-4.9	56.9	50.6	42.4	31.6
5-9.9	14.9	14.4	31.1	25.1
10-19.9	3.4	6.2	13.6	21.7
>2	0.1	2.0	6.5	18.4
All	100.0	100.0	100.0	100.0

Source: *Censuses of agriculture,* 1961 and 1991, Athens, NSSG

Table 2.3
Percentage distribution of holdings by size class

Size class (ha)		Greece			EU12
	Mountainous	Less-favored	Other areas	All	
<2	33.7	39.2	47.7	41.3	37.1
2-<5	31.5	35.0	30.9	32.5	21.7
5-<10	20.4	16.0	13.6	16.1	12.8
10-<20	10.4	6.3	5.8	7.0	10.3
>20	4.0	3.5	2.0	3.1	18.1
Total	100.0	100.0	100.0	100.0	100.0

Source: *Farm structure survey*, 1993, Luxembourg, Eurostat

The proportion of holdings in the small size classes (less than 2 ha) is significantly lower in mountainous areas (65%), than in the less favored (75%) or the other areas of the country (79%). This should be attributed first to the inclusion in agricultural areas in range land and meadows which are mostly found in mountainous areas, and second to the lower percentages of pluriactive farmers that tend to cultivate smaller farms. The structure of livestock producing holdings is presented in table 2.4.

Table 2.4
The evolution of the structure of livestock producing holdings in Greece

Livestock type	1977	1985	1993
Dairy cows			
Number of holdings	131.5[a]	66.2	30.6
Number of animals	559.0[a]	222.0	171.4
Animals per holding	4.2	3.3	5.6
Bovine			
Number of holdings	179.3	93.9	45.0
Number of animals	927.0	674.3	500.9
Animals per holding	5.2	7.2	11.1
Pigs			
Number of holdings	94.0	102.2	54.6
Number of animals	903.7	1,023.2	807.7
Animals per holding	9.6	10.0	14.8
Sheep			
Number of holdings	212.2	188.7	151.0
Number of animals	8,619.1	7,214.6	7,716.5
Animals per holding			
Goat			
Number of holdings	350.8[b]	319.9	192.8
Number of animals	4,741.8[b]	4,382.3	4,901.1
Animals per holding			
Poultry			
Number of holdings	738.0	644.6	422.3
Number of animals	29,734.0	28,160.0	31,850.0
Animals per holding			

a = data for number of households and animals is in thousands
b = figures refer to 1979

Source: *Farm structure surveys*, 1977-93, Luxembourg, Eurostat

The number of holdings producing bovine and dairy cows has declined significantly and so has the animal stock. However the mean herd size has increased and in the case of bovine animals has more than doubled.

The whole sector of beef production in Greece is characterized by eminent structural problems that have been attributed to the lack of appropriate pastures, the low level of investments, the high cost of feedstuffs and the competition from other beef producing European Union countries, despite the geographic position of the country within the European Union that increases transportation costs for imported meat (Lianos and Katranidis, 1993). Both the number of holdings raising pigs and the stock increased up to the mid 1980s and then decreased sharply, while the mean herd size is increasing constantly. The number of holdings with sheep and goats is continuously decreasing while the respective stock was decreasing up to the late 1980s and then started increasing again, although the mean herd size is increasing in both types of livestock enterprises.

Examination of the farm structure in purely area terms is not always the best way to describe agricultural structure changes as other factors of production such as labor and capital are ignored. Farm structure measures in terms of output, income or employment are undoubtedly more useful, but less readily available in agricultural statistics. As described previously, Eurostat calculates an economic size for European Union farm holdings and creates a farm structure typology, based on this measure of gross margin. The Greek mean farm size in economic size units (1 ESU= 1,200 ECUs in 1993), has almost doubled in the period 1983-93 from 3.5 to 6.2, but is still the second lowest, after Portugal, in the European Union. However, the European Union average ESU, for the 12 member countries, is 14.3 and the deviation of Greece from the rest of the European agricultural economy is less than the one measured in area terms.

The share of farms in small economic size units (size of less than 4 ESU) has decreased in Greece since 1980, while the share of holdings in larger size categories has significantly increased, especially the category of over 16 ESUs has increased its share threefold. Compared to the European Union of 12, Greece has less farms in the smallest and largest size classes and more in the medium size classes (table 2.5). Table 2.6 does not exactly confirm the findings of table 1.1 relating farm size in area terms to the age of the farm operator. Farm holders of less than 35 years of age maintain proportionally less small sized and more large size holdings than all other age classes. It is interesting to note that there is an almost linear relationship in the percentage share of the age classes in the large size holdings implying that the younger the age class the higher its share in large economic size units. The joint examination of tables 1.1 and 2.6 points to the conclusion that younger holders operate more intensive farms and, in general, achieve better economic results than older farmers.

Table 2.5

Percentage distribution of number of holdings by economic size in Greece and the EU

Size classes (ESU)	Greece % share of each class		EU12 % share of each class 1993
	1979/80	1993	
< 2 ESU	47.8	32.7	37.1
2 - 3.99 ESU	21.5	20.3	16.4
4 - 7.99 ESU	18.7	22.4	14.3
8 - 15.99 ESU	9.2	16.2	11.6
>16 ESU	2.8	8.4	20.6
All farms	100.0	100.0	100.0

Source: *Farm structure surveys*, 1979/80 and 1993, Luxembourg, Eurostat

Table 2.6

Percentage distribution of number of holdings by economic size and age class of holder

Size classes (ESU)	Age classes of holder				
	< 35	35-<45	45-<55	55-<65	>65
< 2 ESU	25.5	29.3	26.4	29.4	43.8
2 - 3.99 ESU	19.4	18.4	18.8	20.4	22.4
4 - 7.99 ESU	23.6	22.7	23.4	24.4	19.6
8 - 15.99 ESU	19.5	18.3	19.9	17.5	10.5
>16 ESU	12.0	11.3	11.5	8.3	3.7
All farms	100.0	100.0	100.0	100.0	100.0

Source: *Farm structure survey*, 1993, Luxembourg, Eurostat

The land tenure system in Greece includes the tenure arrangements of full ownership of land in which all land used is owned by the farm operator, the partial ownership of land in which the land used is partly rented and partly owned by the farm operator and tenancy where the farm operator rents all the land under utilization. The two dominant types of tenancy in Greece are cash renting and crop sharing. Systems of farm operation under cooperative production and manager operation systems in which managers undertake the operation of farms under management contracts, are less important in Greek agriculture. The importance of tenancy is increasing in Greek agriculture. The percentage of land under tenancy, share farming and other systems except ownership, has more than tripled from 7% of total agricultural area in 1950 to 24.8% in 1993. Lianos and Parliarou (1987) have pointed out that in post-war Greek agriculture, farm land operated by full owners and sharecroppers has declined in favor of land renting which takes place more by larger farms whereas smaller farms occupy a small percentage of total rented land.

One severe structural problem in Greek agriculture relates to the extent of farm fragmentation. Land fragmentation is rooted in the early land reform and land redistribution policies followed by the Greek state in the late 19th and early 20th century. The land distributed to each farmer consisted of 4 to 6 plots at different distances from the farm due to the insistence of the farmers on receiving equal shares of land of varying qualities together with surveyors' efforts to prolong their role in the proceedings (Thompson, 1963). Farm fragmentation in Greece may be attributed to social, physical and economic factors. Of the social factors, inheritance and the custom of dowry agreements have had profound impacts on fragmentation. Due to the high value placed on land in Greek rural society, tradition commands that all heirs receive equal shares of land while dowry is still in the form of several plots of land and animals. On the other hand, physical diversity of soil fertility, slope and terrain leads farmers to acquire land of widely varying productive potential. Especially in mountainous communities, scattered holdings acquired through dowry and inheritance provide a wide range of environmental niches which assures the hedging of bets on a single crop and helps to provide a wide range of micro-environments suited to a wide range of crops (Forbes, 1976). Thus there are grounds on which piecemeal inheritance and transfer of land through dowry is encouraged in highland Greece while this is not the case in plain communities where the environment is more or less homogenous. Finally, one major force toward fragmentation has been the dependence of small farmers on urban-based creditors that required farmers to sell small pieces of land to pay back the loan when they could not settle their debt otherwise (Andreades, 1906).

Table 2.7 shows basic fragmentation figures and the distribution of farms per number of plots cultivated in Greece for 1978 and 1993. The number of plots per holding has increased while the mean size per plot has remained unchanged. The percentage of farms cultivating highly fragmented holdings (more than 6 plots) has increased between 1978 and 1993. There is a very clear distinction between farms in less favored or mountainous areas and the total of Greek farms regarding fragmentation. Fragmentation in less-favored areas is extreme with holdings cultivating a mean of 8.1 plots, while almost 50% of the holdings cultivate farms that comprise of more than 6 plots.

Fragmentation has not been resolved by consolidation projects started in 1948 and it continues up to this day. To date, almost 25% of agricultural land in Greece has been consolidated by compulsory and voluntary consolidation schemes (Keeler and Skuras, 1990).

Table 2.7
Farm fragmentation in Greek agriculture 1978, 1993

	1978 All	All	1993 Mountainous	Less-favored
Basic Data				
Mean farm size (ha)	3.6	3.9	3.5	5.1
Mean number of plots per holding	6.1	6.4	7.2	8.1
Mean size per plot (ha)	0.6	0.6	0.5	0.6
% distribution of farms per number of plots				
1-2	27.3	26.0	19.2	20.2
3-5	35.8	34.3	34.5	30.1
6-9	19.6	20.9	23.2	22.8
10-14	9.3	10.4	12.5	13.3
>14	8.0	8.4	10.6	13.6

Source: *Farm structure surveys 1977-93*, Luxembourg, Eurostat

Case study research on consolidation in a typical semi-mountainous commune in Greece showed that the intense fragmentation accounting for 40.8 parcels per household in 1961 reduced to 4.6 in 1981 while the mean size increased from 0.23 ha in 1961 to 1.61 ha in 1981. Consolidation in this commune was followed by increased productivity. From an environmental point of view, consolidation in Greece has been held responsible for the uprooting of natural farm borders, destruction of physical features and traditional rural landscapes, soil erosion and the elimination of the natural habitat for many fauna species.

Finally, one should note that farm structure is highly differentiated among regions. In the regions of Attica, Epirus, Central Greece and the Aegean Islands, the share of farms that are less than 2 ESUs exceeds 40%. On the other hand, in the regions of Eastern and Central Macedonia, Peloponnese and Thessaly, the percentage of farms that are over the 4 ESUs size limit exceeds 50%.

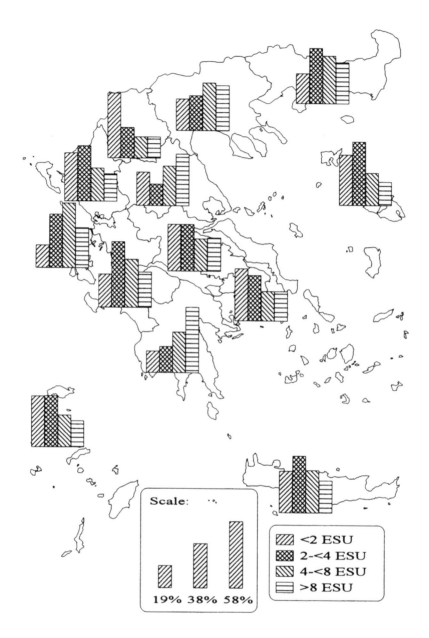

Map 2.1: **Percentage regional distribution of holdings by economic size units**

Source: *Farm structure*, 1993, Luxembourg, Eurostat

These findings confirm earlier research relating the regional distribution of holdings to farm size classes to factors such as the degree of industrialization, agricultural mechanization and to population pressure on agricultural resources (Lianos and Parliarou, 1986).

Land use

Agriculture represents the most extensive land use of the country and together with pastures accounts for almost 70% of the total area. Forests account for 22% and urban land for only 4%, one of the lowest figures among European Union member countries. No spectacular changes concerning the share of the major land uses have occurred in last 30 years. The percentage of non-utilized agricultural land and pastures, accounts for almost 26% in mountainous areas and for only 3.7% in the rest of the country. Land abandonment is a major issue in Greek agriculture. The abandonment of marginal agricultural and range land has been very high, especially in mountainous and less favored areas. Farmers abandon land which is marginal in terms of production capabilities or location and which is also difficult to sell or rent out. Land abandonment is the result of rural desertion due to internal and external migration. In 1983 abandoned land was estimated to account for almost a half million hectares, 60% higher than the respective 1974 figure resulting in a yearly rate of 16,000 ha in the decade from 1974 to 1983 (Tsoumas and Tasioulas, 1986). Taking into account that farmers over 55 years old cultivate more than 50% of the total utilized agricultural land in Greece, together with the low rate of succession at least in mountainous and less favored areas, the rate of abandonment will increase in the future. Abandoned land is comprised mainly of pastures and ranges (45%), arable land and plantations (36%), forest (18%) and other non-usable areas. Almost 50% of abandoned land is in mountainous areas, 37% in semi-mountainous and 15% in plain areas (Tsoumas and Tasioulas, 1986). Almost 63% of abandoned land is owned by people that have left their villages and thus, it is not surprising that the geographic distribution of land abandonment coincides with the distribution of rural out migration.

The pattern of agricultural land use by type of cultivation has changed in response to production changes in demand in the last 35 years (table 2.8). The area under cereals has been constantly declining but this should not be attributed to any structural policy measures that were successful in reducing the land cultivated with cereals in other European Union countries. Schemes aiming to reduce cereal surpluses, known as land set-aside programs, were first introduced in the late 1980s and were not successful in attracting many farmers in Greece due to, mainly, low payments and small farm size. The land cultivated with edible pulses and fodder seeds has experienced a dramatic fall especially after 1981 due to falling prices and competition from other European Union member countries. Area under industrial plants has been declining over the period 1961/81 but increased substantially after 1981 due to the high increases in land cultivated with cotton that grew almost four times in the respective period while land under tobacco slightly decreased due to the Common Organization of the Markets that introduced a quota scheme in 1992. Fodder plants and vegetables exhibit a fluctuating stabilization, while melons, watermelons and potatoes show a slight decrease.

Table 2.8
Distribution of utilized agricultural area among basic cultivations

Cultivation	1961	1971	1981	1996	% 61/96	% 81/96
Cereals	17,565.6	16,294.7	15,884.3	13,179.4	-25.0	-17.0
Edible Pulse	1,606.1	828.4	504.8	184.2	-88.5	-63.5
Fodder seeds	999.9	492.9	208.6	96.8	-90.3	-53.6
Industrial plants	3,646.1	2,843.8	2,801.5	5,503.5	50.9	96.5
Fodder plants	2,677.2	3,263.2	2,926.1	3,064.3	14.5	4.7
Melons and potatoes	905.7	839.2	892.4	782.6	-13.6	-12.3
Vegetables	1,086.3	1,142.7	1,220.3	1,204.0	10.8	-1.3
Vines	2,471.3	2,198.3	1,855.8	1,466.8*	-40.6	-21.0
Citrus trees	345.4	434.9	483.7	574.3*	66.3	18.7
Fruit trees	342.5	520.3	663.0	846.3*	147.1	27.6
Dried fruit trees	433.9	566.8	774.2	620.3*	42.9	-19.9
Olive trees	4,163.8	5,349.4	6,357.7	7,127.5*	71.2	12.1
Other trees	22.0	18.4	24.2	67.3*	205.9	178.1
Total	36,365.8	34,793.0	34,596.1	34,717.3	-4.5	0.3

* figures refer to 1992

Source: *Agricultural statistics of Greece 1961-96*, Athens, NSSG

The area under vineyards has decreased as a response to falling prices for low quality wine and the reduced exports for traditional Greek varieties of grapes such as currants and sultans. Vineyards have undergone a thorough reorganization since the introduction of Community schemes for the permanent abandonment (grubbing premium) and the restrictions on planting introduced in 1980 which affected almost 10% of the country's total area under vineyards. The area under citrus and fruit trees and olive trees has increased in the period 1961-96 with the greatest part of this increase occurring in the period prior to 1981. In general, farmers have responded to both policy and market changes and as is shown later in this chapter, to changes in the use of factors of production and especially farm mechanization and increasing yields, by adapting cultivated land.

Rural settlements

Rural Greece has evolved through numerous causal factors, primarily historic, that led to the nucleation of the rural population. There is not any published work that attempts to formally define the physical, economic and/or social rural space in Greece, while the only attempt to a typology or classification of rural areas is very recent (Dimara and Skuras, 1996) and has been described in the first part of this section. It may be argued that the presence of agriculture and the location/remoteness factors are the key elements in defining rural space in Greece. The presence, however, of other resources and land uses or some distinct characteristics of the economic and social life such as depopulation and the structure of settlements should not be ignored. Perceptions of what is rural differ enormously among various categories of the Greek population from the one extreme of the arcadian ideal to the radical Marxist perceptions of rural problems and countryside conflicts. For statistical purposes, the Census of Population of the National Statistical Service of Greece defines the rural population as the population of those municipalities and communes in which the largest locality has less than 2,000 inhabitants, except those belonging to an urban agglomeration. Despite the predominance given to the size of the commune by this definition, it is successful in summarizing the essential rural characteristics namely, small size of the economy, dependence on agriculture specific culture (attitudes, traditions, level of education) and landscape (Carabatsou-Pachaki, 1993). According to the 1991 Census of Population, the rural space accounts for 28.4% of the population while the same figure was 47.5% in 1961. Rural population is structured in more than 5,000 communes that are made up by more than 12,000 localities. Population structure in rural areas is "top heavy" with the elderly. People with an age of over 55 account for almost 30% in rural areas as opposed to less than 20% in urban areas. The rural economy depends heavily on agriculture while other economic activities are not well developed or absent. Almost 65% of the economically active population in rural areas is engaged in agriculture with manufacturing coming second with 8% and commerce and tourism in fourth place with almost 6%.

A number of significant problems act as barriers to the development of rural Greece. Among these problems are the poor quality of life that is due to both the poor quality and level of public services (medical care, education, transport, etc.) and the inadequate infrastructure, the low skills of the work force, the poor quality of administration, the absence of local synergies and extensive land use conflicts. Unfortunately, there are no statistical data specifically referring to rural regions but, a comparison may be drawn between predominately rural prefectures and prefectures dominated by at least one major urban center characterized as predominately urban. Such comparisons have shown that the gap between rural and urban areas for a number of welfare indices is very wide (Carabatsou-Pachaki, 1993). Greece has not institutionalized an integrated rural development policy and rural development is still the object of sectoral policies or appropriate European Union policies designed to promote a bottom-up development pattern. European Union and national policies have attempted to increase the degree of public investments in the sector of basic infrastructure and increase the level of public service provision in rural areas. One main economic objective of rural policy should be the diversification of economic activities with the promotion of enterprises in the processing, marketing and service sector. Another objective is the assistance for the development of non-farm enterprises on agricultural holdings.

Institutions for agricultural and rural development

Institutions and organizations in rural Greece

Organizations and institutions have been viewed from various perspectives, including sociology, economics, political science and behavioral psychology. For the purposes of this section, when reference is made to an organization or an institution (both terms are used interchangeably), the following definition applies: an institution is a consciously and deliberately formed social entity, whose creation can generally be treated to a particular point in time. Furthermore, an institution is a coordinated and structured social interaction made up of people, who through communication, form an internal interaction network or a social grouping which has an identifiable boundary. Institutions are developed to achieve a specific goal which should be interpreted as representing the organization's intentions at a particular time. Institutions and organizations take actions to extend their lives beyond the initial goals established or beyond the lives of their founders (Cawley et al., 1992).

Institutions are bound together by coordination and authority, which may be legal, traditional or charismatic, and by bureaucratic structure. All forms of social and economic activity are influenced by organizational structures of various kinds and a proper understanding of the former requires that the structures within which they operate are taken into account. Organizations

76

have an internal environment, i.e. their internal structures, and an external environment, i.e. the socio-cultural, economic and political structure outside the boundary of the organization. These two environments can provide both opportunities and constraints for the organization acting in rural development and the rural space. Organizations adopt particular strategies in coping with their external environments, including competition, cooperation, co-opting, coalition, etc., implying that they are not passive recipients of external influences (Cawley et al., 1992).

The main rural development institutions in Greece include the Ministry of National Economy, the Ministry of Agriculture, agricultural cooperatives and various semi-independent or independent bodies. The term "rural development" is not clearly defined in the Greek scientific literature or in official documents. It is very difficult to draw conclusions concerning the objectives of rural development policy. On the other hand, regional development policy, which is very often used to offset the lack of rural development policy, has some broadly defined objectives including the reduction of internal migration, maintenance of acceptable employment levels and improvement in the standards of living (CPER, 1980). The major actor in Greek rural development has been the state, through its departments of central government and state-sponsored agencies. The general responsibility for the application of regional development lies with the Ministry of National Economy, while many aspects and issues of rural development are dealt by the Ministry of Agriculture. Over time, responsibility for special aspects of development, not only in rural areas, has been transferred to specially constituted agencies operating at the national level with regional and local offices (Skuras, 1994).

The Ministry of National Economy has the overall responsibility for the coordination and application of European Union and national policies at regional and local levels. The strategies and actions have been variously realized by the Commission of the EU, through different sectoral policies and initiatives, to a European wide rural development policy. The Commission of the EU understood early on that all rural areas do not face the same development opportunities and constraints and are not in need of the same set of measures. The reform of the three structural funds, i.e., the European Regional Development Fund (ERDF), European Agricultural Guidance and Guarantee Fund (EAGGF), and the European Social Fund (ESF) in 1988, defined five priority areas (objectives), among which the Objective 1 areas are relevant to rural development policy in Greece. Objective 1 areas include regions whose development is lagging behind and are in need of development and structural adjustment. The regions in Objective 1 were defined as those in which the per capita Gross Domestic Product (GDP) was less than the 75% of the Community average and cover the whole of Greece, Portugal and Ireland, most of Spain, the south of Italy, Sardinia, Corsica and the French overseas departments. In Objective 1 areas EAGGF finances improvements in rural infrastructure, irrigation and other agricultural plans. The ERDF finances investments for job creation, infrastructure, support to local development initiatives and the ESF supports training activities.

Each Objective 1 region draws up a project for regional development that includes a very strong agricultural and rural development element, usually financed by the EAGGF. These regional development plans are called Regional Operational Programs (ROPs). ROPs are multi funded and sometimes are called Multifund Operational Programs (MOPs) as opposed to single funded national operational programs (OPs). By the end of 1995, the Commission had approved 48 operational programs for implementing the Community Support Framework in Objective 1 regions. Regulation 2081/93 (Commission, 1996) amended older legislation on the reform of the structural funds and widened the scope of the EAGGF into fields such as improvement of rural living conditions, the renovation of villages, a policy for quality products and product promotion, support for applied research and others. The measures included in each Regional Operational Program are varied and are adapted to meet the specific conditions and production systems of each individual region. The measures aim to meet local needs in the fields of rural infrastructure, crop conversion, improvements to production conditions, afforestation, etc. An extended reference to the impacts of the application of the Community Support Framework (CSF) on Greek agriculture is attempted in chapter four.

At a national level, the Ministry of National Economy is responsible for the design and implementation of the Greek regional development policy framework. The regional policy framework consists of a number of measures starting with the first integrated regional development framework established in 1982 by law 1262/82 and consequently amended, replaced and completed by laws 1892/90 and 2234/94. The country is divided into four zones plus a bordering area and the region of Thrace. Measures are gradually higher from the first development zone to the fourth, bordering areas and the region of Thrace and consist of grant-aid schemes, interest rate subsidization, increased rate of capital depreciation and reduced taxation on profits. The size of financial aid differs not only among zones but also among the different sectors financed.

Investments in practically every economic sector are financed but also investments in non-traditional activities such as the creation of technical support centers, the creation of new products, investments for environmental protection and energy saving, the establishment of new research laboratories and quality control mechanisms are also financed. The Greek regional development support framework had a significant impact on the creation and support of many rural enterprises although regional development policy has been heavily criticized for being biased toward industrial development and for offering instruments limited to fiscal and financial incentives and neglecting all directions of modern regional and rural development policy including flexible combinations of policy tools and technology based initiatives.

The Ministry of Agriculture is responsible for the application of the Common Agricultural Policy, and consequently its rural development component, in Greece. Although it has been argued that the CAP, before the 1992 reform, failed to boost rural and regional development in accordance with trends that existed before its formation (Plascovitis, 1983), new directions in the CAP contain a strong rural development component. Rural development policy under the CAP consists mainly of the so-called "horizontal" and "accompanying" measures and the Community initiative called LEADER. Horizontal measures are all measures applicable on Community territory and their nature is justified by the fact that, since the market organizations affect the whole Community, it is necessary to promote structural adjustment and assist all farms in all regions that need support due to changes in the market situation. Under the horizontal measures, special schemes for farm modernization, the installation of young farmers, support to farmers in the less-favored areas and support to the processing, packaging and marketing of agricultural and forestry products have been launched. Accompanying measures commonly known as "structural stabilizers" have been adopted with an aim to limit the increase in agricultural production but have significant impacts on rural areas, offering new opportunities to development. The accompanying measures include an environmental scheme, an aid scheme for early retirement, forestry measures in agriculture, the set-aside of arable land and the extensification of agricultural production programs.

Finally, the LEADER (Liaison Entre Sections de Developpement de l' Economie Rurale) initiative was launched in 1991 and provided direct assistance to rural development initiatives undertaken by local communities or associations. The Commission itself describes the basic concept of the LEADER initiative as follows: "... Its innovative nature resides in its methodological approach, based on programming and management carried out by those concerned at the local level, development of endogenous resources, integration of projects and the demonstration value of networking the local groups participating in the program". Before launching the second phase of the LEADER initiative (LEADER II), the Commission proposed that rural areas should begin a new stage in looking for fresh solutions to their problems. In all areas involving rural development, this means giving priority actions and investment programs that fulfill criteria such as innovation, transferability and have tangible results. By 1996, 67 multi fund LEADER II programs had been approved by the Commission of which 30 are in Objective 1 areas and 37 are in Objective 5b areas. The Commission also established a network at the European level for rural development measures, with a view to assist and enable the transfer of practical experience and promising innovations throughout Europe. The European Observatory for Innovation and Rural Development (AEIDL) spreads the idea of bottom-up/partnership approach to rural development.

Greece has a very long tradition of strong agricultural cooperatives in the agricultural sector. In 1993, there were 7,138 agricultural cooperatives of every kind in rural areas, with almost one million members (Daouli and Lianos, 1995). Agricultural cooperatives have the highest degree of penetration in rural areas among all rural institutions and organizations. Many agricultural cooperatives own significant infrastructure and production installations and provide a wide range of services (supply of inputs, subsidy delivery, short-term credit, etc.) to the remotest areas of rural Greece. Agricultural cooperatives in Greece suffer from extreme fragmentation and long established state intervention that diverts them from their original goals and targets.

The Agricultural Bank of Greece was established in the 1920s and is the main credit institution in Greek rural areas. The Agricultural Bank of Greece is the organization responsible for the execution of agricultural credit and financial instrument policy, state investment finance and the main consultant to agricultural cooperatives. The Agricultural Bank of Greece does not design its own policy but due to its responsibility for the execution of agricultural policy in rural areas, it is an important part in the policy delivery process. The Agricultural Bank of Greece operates long-term programs for the economic and social development of rural areas by financing various activities such as the installation of migrants returning to rural areas, the installation of young farmers, non-farm activities and by operating pilot and demonstration programs such as a project on rural tourism (Dimou, 1984).

Rural manufacturing and small-scale industrial establishments are mostly financed by the Agricultural Bank of Greece since 1983. Rural manufacturing is an important sector offering employment and contributing to the rural product in many areas in Greece. It has been estimated that there are almost 1,750 wood processing establishments with invested capital that exceeds 7 billion drachmas, a value of product exceeding 8.5 billion drachmas and total employment of around 7,000 full-time workers, together with 1,074 metal processing establishments and many other family businesses of almost all types (Mylonas, 1988).

The range of institutions acting and affecting economic life in rural areas extends to semi-independent bodies such as the Greek National Tourism Organization (GNTO), the Greek Organization for Small and Medium Sized

Manufacturing Enterprises and Handicrafts, the Hellenic Industrial Bank and many other smaller or localized institutions. It should be noted that tourism policy is a major component of rural development policy as long as 80% of the total tourism activity takes place in rural areas and especially on the Greek islands. Its contribution to rural development includes provision of seasonal employment, provision of infrastructure and significantly increased incomes (Logothetis, 1991).

Rural development instruments

Policy is delivered to the rural community within a certain procedure that may or may not satisfy the targeted population. A successful rural development policy should, among others, raise adoption rates among targeted groups and accelerate development processes. The organization of policy delivery is a crucial factor that may attract or deter prospective beneficiaries. The first stage of policy delivery is passing on information and raising awareness about existing alternative rural development policy instruments. The second stage deals with the application procedure and the administrative handling of this application. The final stage refers to the speed of payment of the financial assistance and the time elapsed between the date of application and the date of grant/loan receipt. The organization of policy delivery may be examined from facts and perceptions. How the policy's delivery is perceived by a beneficiary and how the policy's delivery is judged by facts collected from the beneficiary's own experience.

Effective rural development policy offers a range of instruments and influences a wide spectrum of development factors. The instruments currently offered by institutions in rural areas include the traditional schemes of grant aid, innovative instruments under the CAP, training, technical assistance and consultations on farm and business management. Grant aid is offered by Operational Programs (OPs), the Greek regional development framework and the CAP, under the various farm modernization schemes. Farm modernization has been a continuous effort in European agricultural policy since the early 1970s, following the Mansholt Plan. Directive 69/70 set the first scheme for farm modernization followed the Regulation 797/85 and 2328/91 as this was amended by Regulation 2843/94. Farm modernization schemes have been designed to improve the viability and competitiveness of European farms while pursuing the rational development of production focusing on reducing costs, promoting quality, economizing on energy, protecting the environment, advancing working conditions on the farm, promoting hygiene and animal welfare, and promoting farm diversification into non-agricultural activities on the farm (Commission, 1996). Almost 13,000 farms have benefited from farm modernization schemes since 1985 in Greece. Recently, the focus of the farm modernization scheme has been restricted to increase competitiveness and assist farm diversification into non-agricultural activities with the sole aim of reducing agricultural product surpluses. Furthermore, aid to farmers in mountainous and the less-favored areas of the Community is granted with a view to assist these farmers to remain in farming and deal with the adverse economic conditions under which they usually operate. Regulation 2328/91 and other specific measures intend to compensate for the, generally, higher production costs in these areas. Almost 190,000 Greek farms receive compensatory allowances every year under this support framework (Commission, 1996).

Measures designed to assist processing and marketing of agricultural and forestry products are provided under Regulation 866/90 and 867/90 with an aim to organize the way products are placed on the market, rationalize and modernize processing and marketing facilities without increasing their capacity, improve the quality of products and rationalize distribution networks. Measures for assistance in the processing and marketing of agricultural and forestry products have been incorporated into Regional Operational Programs for Objective 1 areas.

A very recent survey assessed the perception of policy delivery as concerns the application procedure, administration and handling of the application and speed of payment in sample of grant-aid beneficiary farm enterprises and rural businesses (Dimara and Skuras, 1997). Each beneficiary was asked to rank each of these procedures into a three grade scale from satisfactory to unsatisfactory. All stages of policy delivery were found satisfactory by the overwhelming majority of beneficiaries in both groups. From the different stages of policy delivery, the speed of payment is the one that attracted the highest unsatisfactory percentages, while the most satisfactory procedure is perceived to be the application stage for rural businesses, followed by the administration and application handling for the same group and the application and handling procedures for farm enterprises (figure 2.2).

An assessment of the importance and availability of development instruments to rural businesses and farm enterprises showed that, besides grant-aid, rural businesses are interested in schemes related to marketing assistance, good quality advice, training facilities, and information on regulations and product standards (Noguera and Stathopoulou, 1997), while farm enterprises are interested in the provision of instruments such as assistance in marketing the produce, provision of extension services especially in the use of modern agricultural technology (machinery, methods of production and use of improved varieties), agricultural training and provision of quality standards (Dimara and Skuras, 1997).

Findings from recent surveys indicate that adding grant aid instruments directed to farm enterprises is high, while adding of financial assistance to rural businesses is low (Skuras and Tzamarias, 1997; Dimara and Skuras, 1996b). However hard to measure, adding is more extensive among farm businesses than among rural enterprises, and within rural businesses is lower among those assisted by national schemes than those assisted by the CAP schemes and lower among medium and small rural enterprises than very small (micro) businesses.

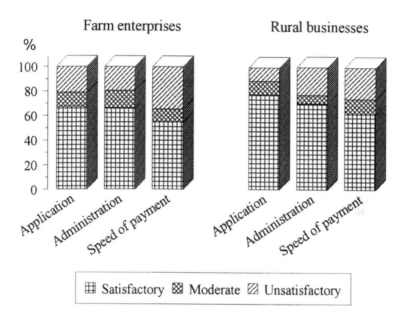

Figure 2.2: **Perception of policy delivery by rural businesses and farm enterprises in Greece**
Source: Dimara, E. and Skuras, D., 1997

The Cork Declaration has clearly argued that: "Rural development policy must be multi-disciplinary in concept, and multi-sectoral in application, with a clear territorial dimension", while, one of the main points in Commissioner Fischler's opening speech at the conference held in Cork dealt with the incoherencies and contradictions among the instruments of policy for rural areas (Fischler, 1996). Important policy documents recognize that agriculture, despite its diminishing importance, will continue to influence important aspects of the rural socio-economic fabric and agricultural policy reforms will give rise to continuing and in many cases countervailing adjustment pressures throughout the agri-food sector (OECD, 1996), while appropriate instruments to help farmers adapt to new challenges are neglected. The use of flexibly tailored combinations of assistance in rural Greece should be increased and complement the use of traditional instruments of grant aid. This approach is, however, associated with certain problems that mainly refer to the extra administration cost involved in using complex multi-instrument sets of support to rural development and lack of human capital and resources able to manage this approach.

Greek institutions operating in rural areas have neither the resources to administer flexible support programs nor the local knowledge and expertise needed to understand the precise types of assistance required in each case. One crucial policy target would be to accelerate capacity-building among institutions in order to design and implement flexible combinations of policy instruments. Institutions may promote a participatory approach which should be based on local and regional initiatives. Furthermore, institutions should enable the local population to articulate their attitudes toward rural development instruments, and design and implement initiatives based on their subjective assessment of availability and need for certain policy mechanisms. Rural development policy should be de-centralized (devolution of rural development policy) in order to become more flexible and selective and match local attitudes and needs. Thus, one crucial policy target would be to accelerate capacity-building among institutions in order to accommodate the design and implementation of flexible combinations of policy instruments in relation to a thorough de-centralization of rural development policy.

Bibliography

Data sources

Bank of Greece (1997), *Monthly Statistical Bulletin* (1960-1997), Economic Research Division, Bank of Greece: Athens.

Commission of the European Communities (1997), *The Agricultural Situation in the Community (1975-1996)*, Office for Official Publications of the European Communities: Luxembourg.

Eurostat (1996), *Agricultural Income*, Office for Official Publications of the European Communities: Luxembourg.

Eurostat (1995), *Total Income of Agricultural Households*, Theme 5, Series D, Office for Official Publications of the European Communities: Luxembourg.

Eurostat (1993), *Farm Structure: 1993 Survey, Main Results*, Theme 5, Series C, Office for Official Publications of the European Communities: Luxembourg.

Eurostat (1990), *Farm Structure: 1989 Survey, Main Results*, Theme 5, Series C, Office for Official Publications of the European Communities: Luxembourg.

Eurostat (1989b), *Manual on Economic Accounts for Agriculture and Forestry*, Theme 5, Series E, Office for Official Publications of the European Communities: Luxembourg.

National Statistical Service of Greece (1997), *Agricultural Statistics of Greece (1960-1996)*, National Statistical Service of Greece: Athens.

National Statistical Service of Greece (1996), *Statistical Yearbook of Greece 1994/95*, National Statistical Service of Greece: Athens.

National Statistical Service of Greece (1995), *Labor Force Survey 1994*, National Statistical Service of Greece: Athens.

National Statistical Service of Greece (1991), *Census of Agriculture*, National Statistical Service of Greece: Athens. (available on electronic media)

National Statistical Service of Greece (1991), *Census of Population*, National Statistical Service of Greece: Athens. (available on electronic media)

European Union legislation - Policy documents

Commission of the European Communities (1997), *Agenda 2000, Volume I and II*, DGVI Home Page (http://europa.eu.int/en/comm/dg06/index.htm).

Commission of the European Communities (1996), *Structural Funds and Cohesion Fund, 1994-1999: Regulations and Commentary*, Office for Official Publications of the European Communities: Luxembourg.

Commission of the European Communities (1995), *Study on Alternative Strategies for the Development of Relations in the Field of Agriculture Between the EU and the Associated Countries with a View to Future Accession of These Countries*, CSE (95) 607, Office for Official Publications of the European Communities: Luxembourg.

The Cork Declaration (1996), *Rural Europe - Future Perspectives*, DGVI Home Page (http://europa.eu.int/en/comm/dg06/index.htm).

References

Andreades, A (1906), 'The Currant Crisis in Greece', *Economic Journal*, Vol. 9, pp. 41-51.

Carabatsou-Pachaki, C. (1993), *Rural Problems and Policy in Greece*, Discussion Paper No. 18, Center for Planning and Economic Research: Athens.

Cawley, M., Gilmor, D. and McDonagh, P. (1992), *Changing Farming Systems: An Approach to Institutional Research Methodology*, Working

Paper 4, Reserach Project: CAMAR (8001-CT91-0119) on Alternative Farming Systems in the Lagging Regions of the EC.

Center of Planning and Economic Research (1980), *Regional Development Plan, 1980-85,* Center of Planning and Economic Research: Athens. (in Greek)

Damianos, D., Demoussis, M., Dimara, E., Fissamber, V., Koniotaki, A. and Skuras, D. (1994), *Regional Report for West-Central Greece*, Working Paper 14, Research Project: CAMAR (8001-CT91-0119) on Alternative Farming Systems in the Lagging Regions of the EC.

Daouli, J. and Lianos, T. (1995), 'The Institutional Economic Framework as a Factor of Development', in Kindis, A. (ed.), *'2004': Greek Economy at the Door-Step of the 21st Century*, Ionian Bank of Greece: Athens.

Dimara, E. and Skuras, D. (1997), *Regional Report for Greece*, Working Paper 29, Research Project: AIR (AIR3-CT94-1545) on The Impact of Public Institutions on Lagging Rural and Coastal Regions.

Dimara, E. and Skuras, D. (1996), *Microtypology of Rural Desertification in Greece*, Working Paper 20, Research Project: AIR (AIR3-CT94-1545) on The Impact of Public Institutions on Lagging Rural and Coastal Regions.

Dimou, N. (1984), *Agricultural Credit in Greece*, Agricultural Bank of Greece: Thessaloniki. (in Greek)

Fischler, F. (1996), 'Europe and its Rural Areas in the Year 2000: Integrated Rural Development as a Challenge for Policy Making', European Conference on Rural Development, *"Rural Europe - Future Perspectives"*, Cork, 7-9 November.

Forbes, H. A. (1976), 'We Have a Little of Everything: The Ecological Basis of Some Agricultural Practices in Methana, Trizinia', in Dimen, M. and Friedl, E. (eds.), *Regional Variation in Modern Greece and Cyprus: Towards a perspective on the Ethnography of Greece*, Annals of the New York Academy of Sciences: New York.

Keeler, M. and Skuras, D. (1990), 'Land Fragmentation and Consolidation Policies in Greek Agriculture', *Geography*, Vol. 75, No. 1, pp. 73-6.

Lianos, T and Katranidis, S. (1993), 'Modeling the Beef Market of the Greek Economy', *European Review of Agricultural Economics*, Vol. 20, pp. 49-63.

Lianos, T and Parliarou, D. (1987), 'Land Tenure in Greek Agriculture', *Land Economics*, Vol. 63, pp. 237-48.

Lianos, T. and Parliarou, D. (1986), 'Farm Size Structure in Greek Agriculture', *European Review of Agricultural Economics*, Vol. 13, pp. 233-48.

Logothetis, M. (1991), *Tourism Policy and Regional Development*, EBE: Thessaloniki. (in Greek)

Mylonas, A. (1988), *Manufacturing in Rural Areas and its Finance*, Agricultural Bank of Greece: Athens. (in Greek)

Noguera, J. and Stathopoulou, S. (1997), 'Institutions and the Provision of Rural Business Development Schemes: A Greek-Spanish Comparison', Paper Presented at the *17ᵗʰ European Society for Rural Sociology Congress*, Chania, 25-29 August.

OECD (1996), *Agricultural Adjustment and Diversification: Implications for the Rural Economy*. Working Party on Agricultural Policies and Markets, AGR/CA/APM(96)5, OECD: Paris.

Plascasovitis, I. (1983), 'A Critique on the "Study of the Regional Impact of the CAP", *European Review of Agricultural Economics*, Vol. 10, No. 2, pp. 141-50.

Skuras, D. (1994), *Review of Rural Development Policy and Analysis in Greece*, Working Paper 5, Research Project: AIR (AIR3-CT94-1545) on The Impact of Public Institutions on Lagging Rural and Coastal Regions.

Skuras, D. and Tzamarias, N. (1997), 'Job Creation by Assisted Rural Businesses', Paper Presented at the *17ᵗʰ European Society for Rural Sociology Congress*, Chania, 25-29 August.

Thompson, K. (1963), *Farm Fragmentation in Greece*, Research Monograph Series No. 5, Center of Planning and Economic Research: Athens.

Tsoumas, T. and Tasioulas, D. (1986), *Ownership Status and Utilization of Agricultural Land in Greece*, Agricultural Bank of Greece: Athens. (in Greek)

3 Agriculture and the macro-economic environment

Although it is indisputable that sectoral price and trade policies affect the performance of the agricultural sector, the contribution of macroeconomic policies is often underestimated or completely ignored. Nevertheless, the macroeconomic environment can impede, foster, or even promote the agricultural development pursued by sectoral policies. As Timmer et al., (1983) state, "an unfavorable macroeconomic environment can ultimately erode even the best plans for consumption, production, or marketing". Moreover they argue that in a distorted macroeconomic environment of rapid inflation, overvalued exchange rate, subsidized interest rates for preferred creditors, minimum wages for an urban working class elite, and depressed rural incentives, rapid growth of agricultural output is extremely difficult to achieve, while the distribution of earned income is easily skewed. In such a case, investments in irrigation, improvements in agricultural research and extension programs, subsidies on fertilizer and modern seeds "contribute to agricultural growth, but in the constraining environment of distorted macro policies, such programs [do] not provide the basis for long run dynamic growth of rural output and incomes" (Timmer et al., 1983). Killick (1985) also argues that "macroeconomic policies which at least avoid severe deficits in the balance of payments and rapid inflation are to be seen as a necessary condition for the promotion of agricultural growth and rural development".

These comments relate with great precision to the case of Greece, since during the period under examination, the country experienced an identical macroeconomic environment (Maroulis, 1992) that was the outcome of a series of socioeconomic and political events which for the purpose of this presentation begins in 1970.

The beginning of the 1970s found Greece under a severe military dictatorship (1967-74). In 1973, the country encountered its first energy crisis that came to add to the severity of the socioeconomic problems caused by the military regime. A year later, democracy was re-established and the new government adopted an expansionary macroeconomic policy in order to heal the wounds from seven years of social neglect by the military administrations. Throughout the period from 1974 to 1981 administrations followed a combination of expansionary income and budgetary policy and of overvalued national currency (drachma). In response to rising government expenditures aiming at establishing socioeconomic cohesion and at improving household income, private consumption increased (figure 3.1), inflation and the budget deficit rose (figure 3.2), as did the borrowing needs of the public sector. Consequently, a continuously deteriorating current account reflected the economy's lost competitiveness, and larger capital (financial) surpluses depicted increasing inflows of private capital in the form of official borrowing from abroad (figure 3.3).

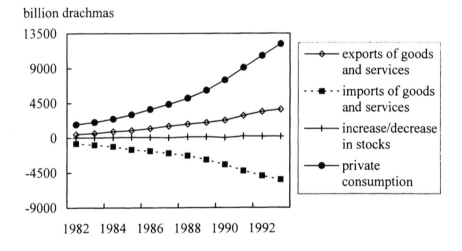

Figure 3.1: **The components of national spending,[*] 1970-93**
Source: *IMF, IFS Yearbook, 1995*

---

[*] The development of government consumption and gross fixed capital formation followed closely the pattern exhibited by exports.

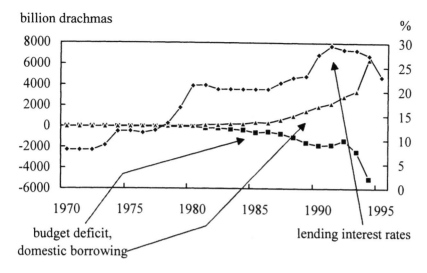

billion drachmas

budget deficit, domestic borrowing

lending interest rates

Figure 3.2: **Budget deficit, domestic borrowing, and lending interest rates, 1970-94**
Source: *IMF, IFS Yearbook,* 1995

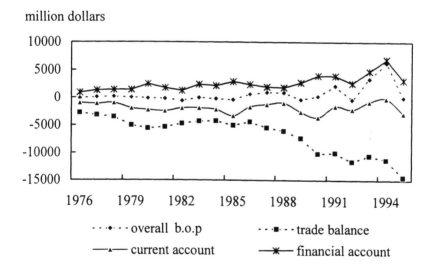

million dollars

- - - ◆ - - - overall b.o.p   - - - ■ - - - trade balance
— ▲ — current account   — ✳ — financial account

Figure 3.3: **The balance of payments (b.o.p) accounts, 1976-95**
Source: *IMF, IFS Yearbook,* 1995

In 1981, PASOK (Pan Hellenic Socialist Party) was elected into power with aggressive social welfare programs. During its first term in office (until the summer of 1985), it followed a much more expansionary fiscal policy and kept interest rates at very low levels. Moreover it kept the drachma overvalued. In 1981, the country ran out of international reserves in its attempt to finance the prevailing account deficit without relying primarily on borrowing. Yet, as can be seen in figure 3.4, the decumulation of international reserves in 1981 did not suffice in impeding the drachma's depreciation. With public debt accumulating and international reserves shrinking to finance it, the country turned to borrowing, which was reflected in increased capital inflows. Nevertheless, debt kept increasing not only in response to larger needs for financing the current account and the budget deficits, but also as a consequence of high interest rates.[1] Greece entered a vicious cycle which made the country less and less competitive and drained it of the resources necessary for economic restructuring.[2]

In 1985 the budget deficit soared, not only in absolute terms, but also as a percentage of GNP (figure 3.5). The country had a payments problem in trying to finance it. This was partly attributed to low levels of savings that resulted from persisting negative real interest rates.[3] Private savings were not enough to finance the 1985 budget and the current account deficits, so the country resorted again to heavy borrowing (figure 3.3) and to further accumulation of public debt. The depletion of international reserves was inadequate to support the value of the drachma, which depreciated by 15% (Zanias and Christou, 1990).

At the outset of the second term in office, the socialist administration introduced a stabilization policy. Although this era lasted only through 1988, it succeeded in giving modest signs of economic recovery. The country curtailed budget and current account deficits (figure 3.2) and ran an overall balance of payments surplus (figure 3.3), which it used to accumulate international reserves (figure 3.4). That improvement in economic performance allowed the country to concede, albeit gradually, to the EEC's demands for liberalization of capital movements.[4] The introduction of (limited) capital mobility in 1987, caused interest rates to rise, in response both to inflationary pressures and the need for financing the budget deficit (figure 3.2).

Unfortunately, during the period from 1989 to 1990 the government resorted again to an expansionary fiscal policy, overvalued currency, and accumulation of debt in response to increasing negative public reaction. Budget deficit skyrocketed and reached a peak in 1990. High interest rates were partly responsible for a soaring debt.

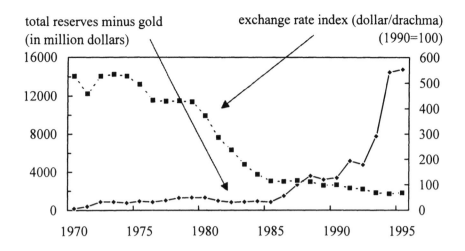

total reserves minus gold (in million dollars) exchange rate index (dollar/drachma) (1990=100)

Figure 3.4: **International reserves and the evolution of the foreign exchange rate, 1970-95**

Source: *IMF, IFS Yearbook,* 1995

Budget deficit as percentage of GNP

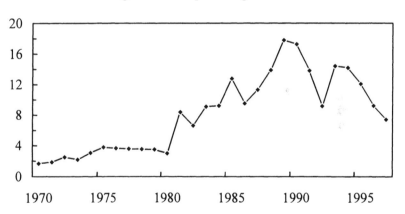

Figure 3.5: **The share of national income accounting for the budget deficit, 1970-97**

Source: 1. 1970-93: *calculated from data taken from IMF, IFS Yearbook,* 1995

2. 1994-97: Ministry of National Economy

From 1991 on, the country followed a stabilization program of tight monetary and disinflationary exchange rate policies, which however did not prevent the competitiveness of the Greek economy from deteriorating (figure 3.6). Modest contraction of inflation rates and of the budget deficit allowed interest rates to take a downward trend. Current account deficit decreased, except in 1992, and reached an all time low in 1993. This enabled the overall balance of payments surplus to rise and to culminate in 1993. Thus, the Central Bank of Greece accumulated international reserves, which permitted the financing of the higher 1992 current account deficit by depleting reserves rather than borrowing. Nevertheless, due to the expenses used to finance the pre-election period of 1993, the budget deficit reached the second to 1989 highest level as a percentage of the GNP.

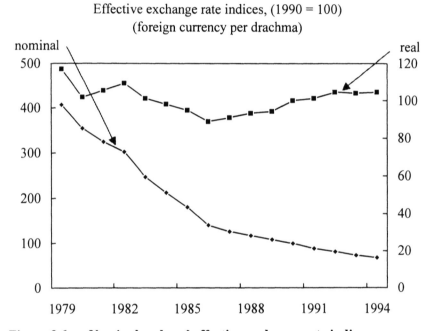

Effective exchange rate indices, (1990 = 100)
(foreign currency per drachma)

Figure 3.6: **Nominal and real effective exchange rate indices (1990=100), 1979-94**

Source: *IMF, IFS Yearbook,* 1995

Thus, Greece has been indeed raising barriers to economic and agricultural development and to structural adjustment by choosing inappropriate macroeconomic policies. In the section that follows, the impact of inflation, overvalued exchange rates, interest rates and budgetary policy will be examined further in relation to agriculture.

Inflation, overvalued currency and green rates

Foreign exchange rates determine the terms of trade between domestic and international markets. As inflation made Greek products relatively more expensive and imported goods more appealing both to domestic and foreign consumers, foreign demand for Greek exports dropped, while Greek demand for foreign imports rose. Since trade balance deteriorated, foreign demand for the national currency (drachma) shrank and the drachma was under pressure to depreciate against the currencies of trade partners. Nevertheless, the (Central) Bank of Greece, operating a managed controlled floating exchange rate regime since the early 1970s, intervened from time to time into the foreign exchange market, in order to protect the national currency from being depreciated, and kept it overvalued.[5] As the full adjustment of foreign exchange parities to free market equilibrium levels essentially offset the impact of inflation, overvaluation implied loss of competitiveness.[6] Whether the result of the relatively higher inflation in Greece or the nominal overvaluation, the real overvaluation of the drachma made Greek products even more expensive for foreigners to consume and impaired the access of Greek exports to foreign markets.

Figure 3.6 depicts the continuous nominal devaluation of the drachma relative to an average of selected foreign currencies (i.e., the nominal effective exchange rate).[7] More importantly, it shows the real effective rate, which compares the competitiveness of Greek products to that of the average of foreign products.

Should domestic prices not change, the drachma's depreciation would imply a rise in Greek exports and a decrease in Greek imports of foreign products. Nevertheless, inflation in Greece rose at increasingly faster rates than in several countries, making Greek products less and less competitive both in domestic and foreign markets. During the period 1981-82 and from 1987 to 1994 the drachma's nominal depreciation was more than neutralized by the relatively higher rate of inflation in Greece, indicating Greece's loss of competitiveness to trade partners.[8] Thus, despite the advantage granted to Greek exports by the depreciating drachma, the cost of Greek products expressed in foreign currency was higher than that of foreign products.

The drachma's overvaluation had the following effects in agriculture (Timmer et al., 1983):

- operated as an implicit tax on agricultural exports by making them more expensive;
- lowered prices of traded products and therefore impaired producers' income, while it indirectly subsidized the consumers of food and other traded products;

- benefited providers of non-traded domestic services (e.g., marketing agents) and suppliers of non-traded goods (e.g., bulky commodities with high transportation costs) since products and services in this category were not exposed to foreign competition and therefore could be priced at higher levels without incurring any losses in market share;
- lowered prices of imported agricultural inputs; therefore effecting
- an ambiguous outcome on the terms of trade in agriculture. Producers of non-traded commodities and consumers of imported inputs will be the beneficiaries at the expense - especially - of producers of traded commodities who use only domestically supplied inputs.

Yet the Greek agricultural sector was not only affected by the rates of the foreign exchange market, but also by the green exchange rates. Every year, the level of Common Agricultural Policy (CAP) institutional prices was decided upon and was denominated in ECUs. The rates used to convert the institutional prices into domestic currencies, however, were not adjusted automatically to the emerging foreign exchange parities, since that would cause turmoil and income instability in agriculture. This adjustment took place gradually by letting the "green rates",[9] that is the rates before any change in foreign exchange's price, approach foreign exchange rates gradually. Decisions on green rate levels were taken every year, together with the decisions for institutional prices.

The mechanism operated as follows: a country with a depreciating currency, like Greece, would not immediately adjust the green rates to the market rates. The domestic currency value of institutional prices was lower than it would have been, had adjustment of the green rates to the market rates taken place. In order, however, to prevent distortion of the common price system, Monetary Compensatory Amounts (MCAs) equal to the difference between the green and the foreign exchange parities were imposed as a tax (negative MCAs) on Greek exports and as a subsidy (positive MCAs) to suppliers of Greek imports (i.e., countries with a relatively appreciating currency). Thus, the flow of Greek exports was hindered, while imports became more competitive in domestic markets.

With Greece's accession to the EEC in 1981, green rates for the drachma equaled foreign exchange rate (figure 3.7). During the period 1981-84 green rates followed the movements of foreign exchange rate. Yet, after the large depreciation of the drachma in 1985 the difference between green rates and foreign exchange rates widened, since consecutive depreciations of the green rate were not enough to catch up with the depreciation of foreign exchange. In 1984, the "switch-over" system was introduced by the European Commission (EC) to mitigate the degree of appreciation of strong currencies. In 1987, the EC began the pursuit of the gradual dismantling of the MCAs through consecutive monetary adjustments (Maraveyas, 1992b; Zanias and Christou, 1990). Taking advantage of these adjustments, Greek green rates approached the foreign exchange level in 1989. Furthermore, until the 1st of February 1995, Greece adjusted its green rates above the drachma's market conversion rates, accounting for the appreciation of the strongest European currency. Thus, another 20.75% of the green rate was added to the drachma's depreciation in attempting to compensate for the effects of inflation. After that date, the green rate mechanism was abolished, green rates were reduced by 20.75% and approached the market conversion rates. The loss from this elimination was compensated by corresponding increases in the institutional price level of the affected products (Commission Européenne, 1996).

Figure 3.7 depicts the green rates for cereals in relation to the drachma's foreign exchange rate.[10] Nevertheless, the movements illustrated are representative.

drachmas per ECU

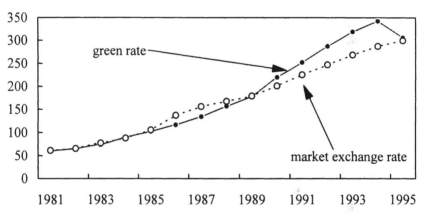

Figure 3.7: **The evolution of green rates and foreign exchange**
rates, 1981-95
Source: *IMF, IFS Yearbook,* 1995 and Ministry of Agriculture

During 1985-89, the green rates were below the market rates, as green rates adjusted very slowly to the depreciating drachma. That implied that the effect of the drachma's depreciation was not passed fully to the prices of supported agricultural products. Producers received lower prices in drachmas, while they faced higher costs of imported inputs. Consequently, agricultural production dropped, agricultural trade balance worsened and foreign exchange revenues decreased. Consumers, paid lower prices for domestic and imported food, which served the anti-inflationary goals of the stabilization policy period. From 1991 to 1995, green rates were allowed to rise above the level of the foreign exchange rate. This development raised the producer price support above the level that would counteract the loss of competitiveness ensued from higher inflation and depicted in the drachma's depreciation.

Ever since the cessation of increases in institutional ECU prices (1984), green rates became a means of manipulating the effect on producers' income to the benefit of farmers. The elimination of this advantage (1995), combined with the international pressure to bring supported prices to world price levels and with the loss of competitiveness, threatened to reduce farmers' income and to accelerate the depopulation of rural areas.

The only way out of this stalemate was a combination of an anti-inflationary macroeconomic environment with a radical restructuring of both the agricultural and rural sectors in ways that will be discussed in the following chapters.

Inflation and land rates

Agriculture was also hit by inflation causing the value of land property to soar. The prolonged period of persistently high general price levels and of negative interest rates made the purchase of land the only sound choice for hedging against inflation. Furthermore, wide fluctuations of real interest rates and lack of alternative investing options for protecting money savings from inflation made people view land more as a means of accumulating and maintaining wealth and less as a factor of production (Demoussis, 1991). Therefore the price of land was not determined by the value of its productivity. In addition, lack of a system for allocating land to various uses and the absence of land banks contributed to the increasing withdrawal of fertile land from agricultural activities. Under these conditions, land renting - the other means of expanding the size of an agricultural enterprise- was hampered both by a legal framework and by a taxing system that minimized mobility of land (Stamatoukos and Spathis, 1991; Maraveyas, 1992a).

Inflation, interest rates and the cost of capital

An effective agricultural credit system is not only instrumental, but also of crucial importance for the success of an agricultural policy. Especially in Greece, where the structure of the agricultural sector poses serious impediments to development, the effectiveness of agricultural policy depends on an agricultural credit system that would avail and mobilize resources for productive expansion of the sector, modernization, and restructuring

(Stamatoukos and Spathis, 1991). Nevertheless, it is imperative that the credit policy should not introduce distortions in the sector.

Subsidization of agricultural credit encourages the adoption of capital intensive techniques, displaces labor, and raises underemployment and unemployment. In addition, subsidized credit introduces distortions in the "informal" private capital markets and widens inequalities in income and wealth. On the other hand, credit subsidization can be the only means to induce small holders to relax their risk-averse attitude and to adopt improved cultivation techniques. Thus, severe credit restrictions should be avoided and priorities for allocating credit in the sector should be drawn carefully (Killick, 1985).

Throughout the post war period, the official agricultural credit operations were performed by the Agricultural Bank of Greece (ATE). Although informal capital markets contributed considerably to financing the needs of the agricultural sector, ATE provided the largest part of total agricultural credit (Sapounas, 1991).

Agricultural credit policy was formed on the basis of social and development criteria (Sapounas, 1991; Stamatoukos and Spathis, 1991). Subsidies to capital and to administratively set interest rates were used as a means of national support to agriculture. Agriculture enjoyed preferential interest rates and loan terms in relation to other sectors of the economy and had priority in funding.[11] The agricultural loan requirements were relaxed and the accommodating terms were very flexible.

Agricultural credit was classified into two broad categories: 1. short term loans, usually with a maturity period of less than a year, aiming at covering the costs of consumable inputs (e.g., fertilizers, feedstuff, labor costs, petroleum, and other cultivation expenditures)[12] and 2. medium-long term loans with maturity in 2 to 30 years, supplied mainly for land reclamation and other permanent improvements, for purchase of livestock, machinery or buildings, as well as for facilitating housing needs (Stamatoukos and Spathis, 1991).

Both credit categories were subsidized so that both types of lending interest rates were significantly lower than the respective rates for other sectors of the economy. This outcome resulted from acknowledging the limited access of agricultural holdings to other sources of capital and the importance of credit for financing operational costs and the costs of an expansion (Stamatoukos and Spathis, 1991).

Yet, credit was very easily supplied even to owners of capital assets, and its coverage expanded beyond the accommodation of real needs (Stamatoukos and Spathis, 1991). On the other hand, credit policy stimulated a behavior toward saving and against self-financing. Furthermore, credit policy led cooperatives and farmers alike to identify agricultural credit with public support. This mentality led to unreliable and irresponsible behavior that was reflected in requests for further credit concessions and in the tendency not to pay off debts in sight of the first income difficulty.

Even worse, as requests for more loans were being satisfied, low liquidity and high debts stopped being accounted for in evaluating petitions for loans and in assessing the financial terms of supplying the loans (Stamatoukos and Spathis, 1991). Thus, the credit policy followed in Greece led to market distortions, irrational use of inputs, wide misallocation of resources, and to widening income disparities.

Since the Agricultural Bank of Greece was not supplying consumer loans, short term loans were often used not only for cultivation purposes, but also for raising spending capacity and household consumption (Stamatoukos and Spathis, 1991). "Short term loans were used to release the household's financial resources for purchasing more consumption goods. Thus, short term loans were not always used to expand the productive capacity of farm holdings, but rather to maintain the family's consumption at a certain level" (Stamatoukos and Spathis, 1991). Consequently, short term loans constituted more than 50% of the loans granted to agriculture until 1993, when for the first time medium-long term loans prevailed (figure 3.8).

Shares of total agricultural loans

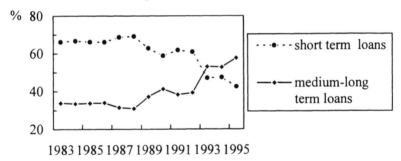

Figure 3.8: **The relative evolution of short term and long term loans to agricuture, 1983-95**

Source: *IMF, IFS Yearbook,* 1995

Varelas and Kaskarelis (1996) argued that, rather than interest rates, liquidity and relative risk constituted the main factors inducing private saving in the economy.[13] In relation to the agricultural sector, Sakellis' (1985) econometric analysis showed that saving behavior was largely determined by the current disposable income and influenced by the inflation rate. Stamatoukos and Spathis (1991) concluded that liquidity and relative yields to capital induced farmers to save. Moreover, they stated that following the credit policy

100

reinforced farmers' reluctance to channel their savings to finance the needs of their enterprises,[14] since it secured the best yield to capital, improved liquidity, and lowered risk.[15] Large enterprises, being able to deposit large amounts of money, particularly benefited by the differences in capital yields. Distortions were such that until 1986, lending interest rates were even lower than the respective deposit rates for savings accounts. This fact stimulated agricultural units to resort to borrowing without really having the need to do so (figure 3.9). Furthermore, in the early 1980s long term interest rates in agriculture were even higher than short term interest rates, thus providing an additional incentive to farmers to borrow for consumption needs rather than for investing in projects that would raise the sector's productivity.

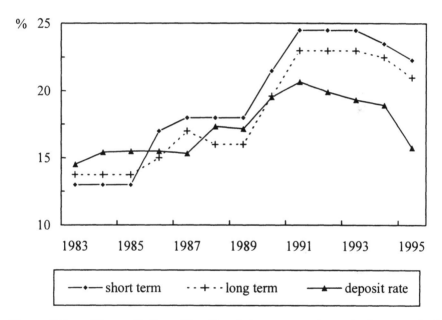

Figure 3.9: **The evolution of lending interest rates in agriculture and of general deposit rates, 1983-95**
Source: *IMF, IFS Yearbook,* 1995, Agricultural Bank of Greece, and Bank of Greece

Until 1993, short term loans in agriculture rose irrespective of the movements of nominal and real interest rates (figure 3.10). Nevertheless, in 1993, when the general lending rates for the economy began to fall and to approach the stabilized lending rates for agriculture, short term loans plunged - in absolute terms - to the 1986 level. As of that time, short term loans began to move in consistency with economic theory.

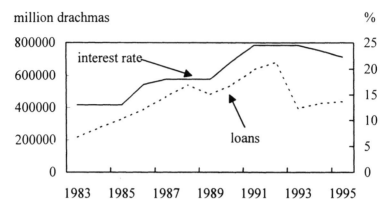

Figure 3.10: Short term lending interest rates and loans to agriculture, 1983-95

Source: Agricultural Bank of Greece and Bank of Greece

From 1983 through 1986, both the real lending rates for agriculture and the real general deposit rate for the economy were negative (figure 3.11). That implied a disincentive to save and an incentive to borrow money either for investing or for spending (figure 3.12). With the introduction of the deregulation of the capital market (1987), real deposit rates remained negative, despite the rise in nominal interest rates ensued by the public sector's borrowing needs and the fall of inflation. Although real lending rates turned positive, the disincentive to save and the low levels of real interest rates effected an increase in short and long term loans to agriculture.

During the period 1988-89, as a consequence to the stabilization policy, nominal lending rates remained stable and inflation fell (figure 3.13), thus providing an incentive to borrow, mainly for investing. Meanwhile real deposit rates turned positive for the first time, thus providing an incentive to save. Although this did not affect long term loans, short term loans dropped in 1989, partly because the expected gains from savings substituted for short term borrowing destined for household consumption. The widening difference between long term and short term interest rates to the benefit of long term loans also contributed to the increase of long term loans relative to the short term ones.

In 1990, falling real lending rates, negative long term rates, and negative real deposit rates underlay a significant rise in long term loans and a modest rise in short term loans. Although in 1991, nominal and real lending interest rates rose, the level of both long and short term loans increased, probably as a result of relatively low levels of real lending rates (3.5% and 5% respectively) and very low real deposit rates (1.2%). After that and through 1995, interest rates either remained stable (1992-93) or decreased. Meanwhile from 1991 through 1995, real deposit rates increased at an increasing rate, while real lending rates became also positive and increasing, due to the needs for financing the deficits.

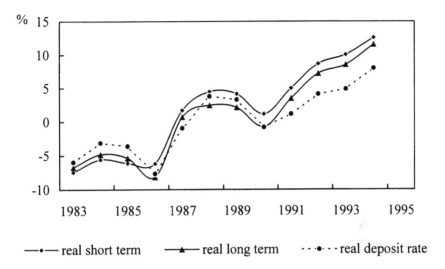

Figure 3.11: **The evolution of real lending interest rates in agriculture and of real general deposit rates, 1983-94**

Source: Approximated by calculating a CPI deflator from *IMF, IFS Yearbook,* 1995 data and deflating data from the Agricultural Bank of Greece and the Bank of Greece

In 1992, inflation retreated significantly, while nominal lending rates were almost constant and nominal deposit rates dropped but at a slower pace. Under these circumstances, short term loans reached a peak and long term loans rose. In 1993, however, real gain from saving reached a little over 5 percentage points, while nominal lending rates remained stable. This development left long term loans unaffected in their upward trend, but convinced farmers to reduce short term borrowing. In 1994 and 1995, nominal lending and deposit rates fell, thus leading to higher loans in either category, although both real lending and real deposit rates were increasing and the distance between real lending and real deposit terms was shrinking.

million drachmas

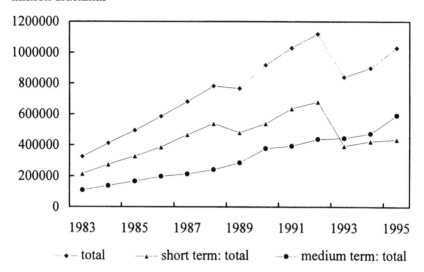

Figure 3.12: **Loans to the private agricultural sector, by major category, 1983-95**

Source: Agricultural Bank of Greece and Bank of Greece

Annual percentage change in CPI

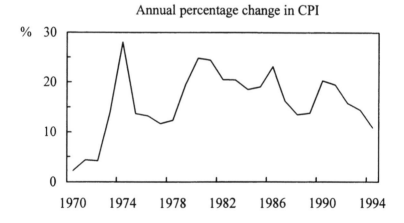

Figure 3.13: **The evolution of inflation, 1970-94**

Source: Approximated by using the CPI index (1990=100) taken from *IMF, IFS Yearbook,* 1995 data

In general, saving, borrowing, and investment behavior of agricultural holdings in Greece did not seem to be particularly responsive to interest rate movements.[16] An explanation could be provided by the farmers' speculative mentality mentioned earlier. Another explanation could be that the agricultural sector was not capital intensive enough, either in activities performed or in techniques used so as to be sensitive to interest rates.[17] In any case, capital subsidization promoted borrowing, rather than investment. That is, interest rate subsidies did prove effective in raising the financial resources available to farmers, but did not provide a strong enough incentive for farmers to engage in investment activities.

Long term loans seemed not to be speculative in nature. The constantly rising levels of long term loans suggested that a few farmers disregarded the interest rate movements and took advantage of the interest rate subsidies to agricultural loans in order to realize their investment plans and raise the productivity level of their enterprise (figure 3.14). Nevertheless, this behavior was not widely adopted, and private investment in agriculture performed poorly.[18] Consequently, rather than stimulating investment, capital subsidization induced higher consumption levels and the excessive use of consumable inputs relative to the level of fixed capital and to the growth rates in investment. Thus, the use of inputs beyond a certain level could not raise productivity; instead it degraded the environment and it wasted valuable resources.

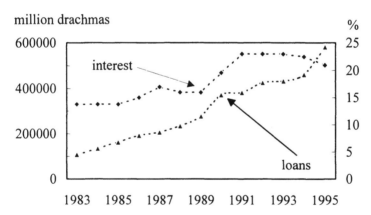

Figure 3.14: Long term lending interest rates and loans to agriculture, 1983-95

Source: Agricultural Bank of Greece and Bank of Greece

Before Greece's accession to the EEC, the use of consumable inputs was promoted both indirectly, by means of interest rate subsidies to short term loans and directly by means of subsidies to consumable inputs. The objectives consisted of inducing producers to higher use of inputs in order to raise productivity and of lowering the cost of production.

Fertilizers, seeds and farm chemicals were subsidized to bring down the cost of production. More specifically, fertilizers were heavily subsidized, in order to compensate for the soaring effect that the two energy crises had on the input's price. On the other hand, subsidies on feed grain aimed at stimulating livestock production, thus raising the degree of self-sufficiency in livestock products and saving foreign exchange. Subsidy levels were such that the cost of inputs facing producers was lower than the cost of producing the inputs (Georgakopoulos, 1991). Specific private investment expenses, including the cost of purchasing agricultural machinery and tools, or the construction cost of greenhouses, were also subsidized (Georgakopoulos, 1991).

As with interest rates, subsidization of consumable inputs effected the irrational allocation of resources. Given the level of investment, the overuse of inputs reduced their productivity and worsened the agricultural trade balance. In response to the subsidies, farmers turned to overmechanization and to more capital intensive techniques of production. Thus, the problem of underemployment in agriculture worsened, environmental attributes were impaired, and agricultural production could not fare any better, as long as investment was not increasing accordingly (Maraveyas, 1992a; Maraveyas, 1989).

Taxation

The agricultural sector received favorable tax treatment, especially until 1996, when a moderate income tax system was introduced in the agricultural sector.

In general, the agricultural sector enjoyed advantages that derived directly from large exemptions on income taxes and indirectly from farmers escaping indirect taxes by consuming self-produced goods.

Nevertheless, taxation has not been used as a means to mobilize land for agricultural purposes. Yet, a tax imposed on non-agricultural land would reduce its value and the costs of production would accelerate restructuring and would raise competitiveness.

Restructuring and terms of trade in agriculture

The changes in the composition of output were effected by the distortions that the CAP introduced in the terms of trade and by the lack of sufficient information on the necessity of restructuring.

Livestock production in Greece specialized in products that were not supported by the CAP (i.e., sheep and goat). Moreover, the production of poultry meat, bovine products, and pig meat was suddenly burdened by a cost that exceeded the cost effected by the adjustment of feedstuff prices (e.g., grain) to the CAP support levels. After the accession, competition lagged due to structural problems. Livestock production was particularly hit, since it necessitated larger investments which were difficult for small farm holders to handle under conditions of financial shortage, economic instability, and rising inflation.

Changes in the terms of trade also affected in part the composition of crop production. In addition, absence of information regarding the long run perspectives for international and domestic tendencies in consumer preferences and in price movements assisted in forming a short run view of the situation that impeded dynamic restructuring.

Disincentives

Partly being the result of inappropriate macroeconomic policies and ineffective sectoral policies and partly being the result of a limited approach to agricultural development, rural areas were stripped of any attractions to rural citizens. Lack of productive and dynamic investment in rural areas effected oversaturation of the labor market, increase in pluriactivity and underemployment in agriculture.

The legal framework was not modified to facilitate the implementation of structural policies, such as the transfer of agricultural holdings to young farmers through the early retirement scheme or the establishment of young farmers in agriculture. Investment in infrastructure did not suffice to revitalize rural areas, reverse depopulating trends, stop demographic aging. Access to information and to incentives for restructuring was not always equally available to all. Participation in decision making was weak. Institutions operated in a non conducive way to agricultural restructuring and to rural development. Sectoral investments proved insufficient for creating jobs and for alleviating inequalities in the distribution of income.

Under these conditions, any attempts to restructure the agricultural sector were destined to effect only short lived improvement in agricultural income, instead of providing the foundation for long term dynamic and sustainable growth.

Notes

[1] Interest rates soared internationally after the second energy crisis (1979), since the resulted inflation raised the cost of money. In Greece, interest rates continued to rise after Greece's full accession to the then European

Economic Community (EEC) in 1981, in response to the country's gradual abandonment of protectionism.

2   The reasons that a country may choose to finance the current account deficit by borrowing rather than by depleting its reserves are two and relate to the fact that international reserves constitute part of a country's monetary base and therefore of the money supply. First, depletion of reserves would imply reduction of the money supply, which could be in conflict with monetary or interest rate goals. Second, the prospect of higher unemployment, which would result from the chain effects of lower money supply, higher interest rates, lower investment and reduced income (GDP) levels, may be deemed unacceptable.

3   Real interest rates were negative from 1973 to 1986.

4   Until that time, Greece had been using a rigid mechanism of restrictions imposed on foreign exchange, in order to hinder the flow of capital out of the country.

5   In doing so, the Bank of Greece decumulated international reserves and distorted the market exchange rate to the extent that it deemed appropriate and feasible. Under a flexible exchange rate regime, the exchange rate adjusts according to the pressure of market forces, there is no need for government intervention and, ceteris paribus, there is no change in the level of official international reserves. Under a fixed regime, the government has to intervene if it wishes to keep the exchange rate at a certain level. Thus, it supplies international reserves when the country runs an overall deficit and accumulates international reserves, when it runs an overall surplus. Yet, there are two ways to finance the current account deficit; either by depleting international reserves or by effecting capital inflows. The latter can occur either by providing incentives to foreigners to invest in domestic assets or by borrowing money. When the level of international reserves is not adequate to finance the current account deficit, full intervention is not feasible and the country has to borrow money (capital inflow). On the other hand, intervention may not be compatible with other macroeconomic goals and therefore may be deemed inappropriate.

6   An exchange rate policy of overvalued currency aims at preventing the cost of consuming imported goods from rising and the general price level (i.e., inflation) from growing.

7   According to Rivera-Batiz and Rivera-Batiz (1985), the selected countries should be major trade partners.

8   Devaluation and depreciation (revaluation and appreciation) define the loss (gain) of value of a national currency relative to foreign currencies. The difference in the terminology depends on whether the acting country

108

follows a fixed (pegged) or a freely flexible (free-floating) exchange rate regime, respectively (Rivera-Batiz and Rivera-Batiz, 1985).

[9] "Green rates" apply only to agricultural products.

[10] In July 1987, Greece began to use different green rates among categories of major agricultural products.

[11] Subsidies to agricultural credit have been financed by the public sector.

[12] Costs of consumable inputs constitute intermediate consumption.

[13] Since the main source of Greek banks for securing capital is private savings, the (opportunity) cost of money is determined by the deposit rate (Stamatoukos and Spathis, 1991).

[14] Greek farmers avoid using their savings to finance agricultural needs. Instead, farmers keep their savings in deposit accounts, or reserve them in anticipation of family emergencies, or avail them for purchasing urban real estate or agricultural land. Thus, Greek farmers have a tendency to finance their farming needs by borrowing.

[15] Rates for depositing savings with the Agricultural Bank of Greece also have been favorable.

[16] Moreover, responses to movements seem to register beyond a high enough level of interest rates and as the gap between general and agricultural credit terms shrinks.

[17] According to Rausser et al. (1986), the more capital intensive farming is the more movements in real interest rates have significant effects on the cost structure of agricultural production. Especially stock carrying of storable products and breeding stock of non-storable commodities (e.g., live cattle and live hogs) are very sensitive to interest rates.

[18] According to Maraveyas (1989) the dual nature of agricultural holdings, operating both as consumers and producers, effected many times the use of investment aids in consumption and other times the use of income support funds in investment activities, thus questioning the effectiveness of investment support in restructuring the agricultural sector.

Bibliography

Commission Européenne (1996), *La Situation de l' Agriculture dans l' Union Européenne*, Rapport 1995, Bruxelles.

Demoussis, M. (1991), 'Impacts of inflation on Greek agriculture', *Nea Oekologia*, No. 21, May. (in Greek)

Georgakopoulos, T. A. (1991), *Greek Agriculture in the EEC Framework: The Impact of Accession*, Agricultural Bank of Greece: Athens.

IMF (1995), *International Financial Statistics Yearbook*, IMF Publication Services, Washington, D.C.

Killick, T. (1985), 'Economic Environment and Agricultural Development: The Importance of Macroeconomic Policy', *Food Policy*, Butterworth & Co. Ltd, February.

Maraveyas, N. (1989), *The Accession of Greece to the European Community: The Effects on the Agricultural Sector*, Foundation for Mediterranean Countries: Athens. (in Greek)

Maraveyas, N. (1992a), *Agricultural Policy and Economic Development in Greece*, Nea Synora: Athens. (in Greek)

Maraveyas, N. (1992b), *The European Integration Process and Greek Agriculture in the 1990s*, Hellenic Center for European Studies: Athens (in Greek).

Maroulis, D. (1992), *Economic Analysis of the Macroeconomic Policy in Greece during the period 1960-1990*, Center of Planning and Economic Research: Athens. (in Greek)

Rausser, G. C., J. A. Chalfant, H. A. Love, and K. G. Stamoulis (1986), 'Macroeconomic Linkages, Taxes, and Subsidies in the U.S. Agricultural Sector', *American Journal of Agricultural Economics*, May.

Rivera-Batiz, F. L. and L. Rivera-Batiz (1985*), International Finance and Open Economy Macroeconomics*, Macmillan: New York.

Sakellis, M. G. (1985), *Saving Behavior of Agricultural Households in Greece*, Agricultural Economics Studies, No. 18, Agricultural Bank of Greece: Athens, March.

Sapounas. G. S. (1991), *Development of the Agricultural Sector: Problems and Perspectives*, Agricultural Economics Studies, No. 42, Agricultural Bank of Greece: Athens. (in Greek)

Stamatoukos G. and P. Spathis (1991), *Agricultural Credit*, Agricultural Bank of Greece: Athens. (in Greek)

Timmer, P. C., W. P. Falcon, and S. R. Pearson (1983), *Food Policy Analysis*, Johns Hopkins University Press: Baltimore.

Varelas, E. and I. Kaskarelis (1996), *Cycles and Economic Policies*, Agricultural Bank of Greece: Athens.

Zanias, G. P. and Christou G. K. (1990), *The Impact of Implementing the Agro-monetary system in Greece: Methodological Approach and Results*, Paper presented at the Panhellenic Conference for Agricultural Economics, Agricultural University of Athens, Athens, November 23-24. (in Greek)

110

4 Agricultural, regional and structural policies

The pre- and post-accession agricultural regional and structural policies

Agricultural policy before accession

Pre-accession agricultural policy in Greece was of major significance, justified by the sector's importance in society and the economy. The post war Greek economy, during the 1950s and the 1960s, was in need of productive agriculture. The principle agricultural policy objectives, as officially stated did not differ from the objectives adopted by the Common Agricultural Policy. The Greek state aimed at increasing productivity by making rational use of resources. They tried to assure a fair standard of living for the rural population by improving agricultural income and stabilizing the domestic market by promoting supply increases in deficient sub-sectors, and at maintaining consumer prices at levels that would not interfere with the general economic situation. In addition the state, by means of policy instruments, aimed at increasing the income gap between rural and urban citizens. This latter goal was deemed necessary as the phenomenon of internal migration and immigration of the rural population to Western Europe, the United States, Canada and Australia was threatening the social fabric of the Greek countryside. In spite of the similarities between the national policy implemented in agriculture and the Common Agricultural Policy during the pre-accession period, certain major differences existed. In order to identify such differences and compare pre- and post-accession agricultural policy in Greece it is necessary to review the major policy elements.

The increase in productivity was sought by measures affecting entire areas such as the implementation of large scale irrigation and infrastructure projects as well as measures of a micro economic character such as incentives for the purchase of equipment and machinery as well as input subsidies for fertilizers, feedstuffs, seeds and chemicals. The state's effort to enhance and stabilize farmers' income was centered around price policy coupled with surplus purchasing along with direct control over the cultivated area for products such as tobacco and rice (Pepelasis, 1980). Farmers were guaranteed minimum intervention prices for wheat, tobacco, dried raisins and olive oil. Direct income aid was also granted in less favored regions of the country. Consumer prices were kept at low levels by means of consumer subsidies whereas export subsidies were granted in order to bridge the difference between domestic and international prices.

Market stabilization was also pursued by implementing conventional agricultural policy measures at the border such as customs duties and levies, import control and even prohibition of imports for sensitive products during certain periods of the year. Agricultural loans for operating capital in the short run as well as long run improvements were granted by the Agricultural Bank of Greece at heavily subsidized rates. Farmers were also favored by the fiscal policy implemented as they were exempt from paying income tax.

The panoply of policy instruments was often erratic, inefficient and contradictory. Price and income support measures revoked any effort for crop restructuring. Short run income relief measures clashed with measures intended to raise productivity. It was apparent that a number of effective national pre-accession policy measures, such as input subsidies or duties against imports for agricultural products of Community origin would have to be eliminated upon the country's accession to the European Economic Community as being incompatible with the Common Agricultural Policy. However, policy reforms were to take place gradually over a five to eight year period until the transitional period agreed upon in the context of the Treaty of Accession was over.

More specifically the national pre-accession policy for the more important sub-sectors of Greek agriculture can be outlined as follows: in the cereals sub-sector, national policy was based on producers' price support. The national processing industries were subsidized in order to be induced to purchase the largest possible quantities. A Cooperative Association for the Management of Agricultural Produce (KYDEP) was assigned intervention responsibilities by the state making the system guarantee minimum prices for cereal farmers in Greece. International trade for cereals was also fully controlled by the state that could impose quantitative restrictions or issue import and export licenses. In the second half of the 1970s, the last decade before accession, this intervention mechanism administered roughly one third of the entire national production, influencing the average market price which normally settled around the level of the intervention price. Shortly before accession, national prices were somewhat lower than Community intervention prices except for high quality corn produced in Greece whose price was higher than the Community's price. Durum wheat producers in mountainous and less favored areas received more substantial income aid than that given by the Community. Before accession, Greece and the Community as a whole were surplus producers of common wheat whereas in the case of durum wheat Greek surpluses could be disposed of on the Community deficient market. The national market for fresh fruit and vegetables was thoroughly protected against imports before accession. Foreign competition was eliminated by means of import restrictions and controls. The state imposed a minimum price that had to be respected by processors and exporters alike. In the beginning of every marketing year the level of export subsidies was fixed by the state.

Public policy in the olive oil sector was geared toward protecting farmers' incomes as well as encouraging exports. Domestic market organization relied upon a system of a guaranteed price for the producer. The state, by means of Eleourgiki, an olive oil cooperative association, would offer the possibility of unlimited intervention at a fixed price. National policy for cotton was carried out by means of KYDEP which would buy, on behalf of the state, any excess quantities at a price corresponding to the international price of the produce. Any farmer could bring in any quantities of cotton that were not sold in the market at a price that he could consider satisfactory.

The domestic market organization for wine was not centralized before accession. Private and cooperative firms would receive state aid, provided they paid a minimum price to the producers during periods of excess supply and low market prices. The state also provided for an export subsidy and strictly controlled imports by issuing import licenses and imposing import duties accordingly.

State control concerning the production of tobacco was strict and was implemented by a system of licenses that was administered to all farmers. The National Tobacco Organization was responsible for the evaluation and classification of the produce on the basis of its quality and intervention prices were set accordingly. The Organization would buy all surplus quantities at a minimum price on behalf of the state.

The state's main preoccupation in the highly deficient beef meat sector was to maintain consumer prices at the lowest possible level. A ceiling price not to be exceeded by retailers was fixed and marketing margins were monitored. There were no minimum guaranteed producer prices in force whereas, at times, a consumer's subsidy would be available by the state. No specific market organization was provided for the sheep and goat meat sub-sector and prices determined by supply and demand. The state did not impose a ceiling price as it did in the case of beef meat. Strict import controls were imposed taking into account the situation concerning domestic demand.

Investment aids were provided to farms along with state aids for the purchase of animals for breeding purposes. Finally, the market for pig meat was organized in a similar way to that of beef meat by fixing a ceiling price and monitoring marketing margins. Heavy state subsidies were granted up to the year 1974 for the development, extension and relocation of capital intensive production units.

National policy did not incorporate effective policy instruments that would enhance and rationalize marketing systems and domestic production. Marketing was not sufficiently linked to contemporary market conditions. Any efforts made to improve product quality and increase productivity did not bring about any concrete results.

Greece and the Common Agricultural Policy

Greece had been associated with the European Economic Community since 1962 on the basis of an Agreement that was interrupted on April 21, 1967 by a military coup. Everything was at a standstill until the fall of the dictatorship in 1974.

Agriculture had been a major area of concern both for Greece and the Community. From the national point of view, the concern expressed was justified by the sector's importance in relation to the rest of the economy (Pepelasis, 1980). From the Community's point of view the country's accession was linked to the enlargement of Spain and Portugal. At the time, the Common Agricultural Policy being implemented throughout the Community of Nine was causing the precipitation of a major crisis due to the pressures exercised on the Community's budget. Against a background of relative economic stagnation it was becoming evident that changes in agricultural policy had become pressing and inevitable. In spite of adjustments already adopted to make agricultural policy tolerable in order to accommodate Italy, the Common Agricultural Policy was designed according to the needs of northern countries by placing emphasis on protecting the incomes of milk, meat and cereal producers. Tensions were mounting as very large income transfers took place from poorer to richer countries and regions, primarily through trade. It was doubtful whether the Common Agricultural Policy could deal with the central issue of income distribution raised by Greece's accession.

For political reasons, Greece hastened to join (1981) the European Union before Spain and Portugal (1986) on the basis of an Agreement reached in December 1978, even though it was unprepared. For cultural and political reasons Greece was welcome by the European Community as its tenth member state. Two major principles prevailed: 1. the acceptance by the new member state of all community rules and regulations (acquis communautaire) in their current form as provided by treaties and secondary legislation. Adherence to this principle practically meant that any problem of adjustment encountered would not be solved by changes in these rules; and 2. measures implemented throughout a five to seven years transitional period would ensure a balance of reciprocal advantages (Pepelasis, 1980). In agriculture, Greece undertook the obligation to progressively eliminate all customs duties against agricultural product imports of community origin by adopting the Common Customs Tariffs. This would guarantee a gradual alignment of producer prices to those of the Community as well as to eliminate all national subsidies by adopting the Common Agricultural Policy instruments.

Although changes in land, labor and capital in European agriculture, as a consequence of Greece's accession was very modest, the Community was unwilling to grant any special provisions as they would be recognized as a precedent by Portugal and Spain whose respective dimensions were by far very important. Greece's accession led to an increase of total area of the Community by 9%, arable land by 6%, population by 4%, agricultural population by 18% and the total number of tractors by 2% (table 4.1). Greece, Portugal and Spain, together, would increase the total EC population by 20% and its farm population by 60%. Regions where agriculture is practically the sole economic activity would be substantially increased and the farm lobby would be strengthened.

Table 4.1
Changes in land, labor and capital in EC agriculture
due to Greece's accession (1976)

	EC9[a]	% Increase
Total land area[a]	150.4	9
Arable area[a]	46.3	6
Total population[b]	258.7	4
Agricultural population[b]	20.3	18
Number of tractors[b]	4.7	2
Number of combine harvesters[b]	0.5	1

[a] in million hectares
[b] in millions

Source: *Production yearbook 1976,* Rome FAO, 1977

There seemed to be no political willingness among the nine member states to raise the CAP spending and sanction larger income transfers to attain greater equity (Marsh, 1980) by granting protection to Mediterranean agriculture. Greece's contribution to the agricultural production of the EC9 was marginal for most products except for tobacco, olive oil, oranges and tomatoes for processing (table 4.2). The accession of Portugal and mainly Spain had more serious implications, as they would cause an increase of production for certain products by roughly 90 to 170%. The expected impact of Greece's accession on trade is presented in table 4.3. Those statistics demonstrate complementarity in products such as rice, barley, olive oil, tobacco, citrus fruit, butter and cheese. Surplus domestic production of wheat and flour would add to the Community's surplus whereas a substantial deficit for beef meat in the domestic market would cause an increase in total Community imports. The impact on the Community's trade patterns however was

Table 4.2
Contribution of Greece to the agricultural production
of the EC (in '000 tons, 1978)

Product	Greek production	Greek production as % of EC9
Wheat	2,660	6
Rice	92	1
Barley	956	2
Maize	537	3
Potatoes	944	2
Fruit	3,054	12
Vegetables	3,481	12
Tomatoes	1,751	34
Grapes	1,375	6
Wine	435	3
Sugar Beet	3,000	4
Oranges	600	42
Tobacco	113	66
Olive Oil	202	52
Beef and Veal	104	2
Sheep and Goat	76	15
Pig Meat	123	1
Milk	705	< 1
Cheese	165	5
Butter	7	< 1
Eggs	119	2

Source: Marsh, J. S., 1980

expected to be intensified after the accession of Portugal and Spain with the exception of tobacco and wheat for which the two prospective new member states were not importers.

Table 4.3
Expected (in 1978) impact of Greece's accession on
EC trade pattern (in million US dollars)

	Trade balance			
	EC9	Greece		Greece as % of EC9
Cereals	-135.4	-10.0	7	(Addition to imports)
Wheat and flour	68.9	72.5	105	(Addition to exports)
Rice	-174.4	2.5	1	(Reduction of imports)
Barley	228.7	-31.0	14	(Reduction of exports)
Maize	-1,407.3	-144.0	10	(Addition to imports)
Sugar	149.4	4.1	2	(Addition to exports)
Olive oil	-127.4	82.5	64	(Reduction of imports)
Oil seeds	-1,973.5	-2.7	0	(Negligible impact)
Wine	432.0	23.8	6	(Addition to exports)
Tobacco	-1,602.9	208.8	13	(Reduction of imports)
Citrus fruit	-882.3	75.7	8	(Reduction of imports)
Cattle	-94.1	-6.1	6	(Addition to imports)
Pig meat	-28.0	0.0	0	(No impact)
Beef and veal	-117.3	-270.0	230	(Addition to imports)
Butter	193.4	-4.5	2	(Reduction of exports)
Cheese	196.1	-5.7	3	(Reduction of exports)

Source: Marsh, J. S., 1980

The Common Agricultural Policy lacked the proper instruments that could be effective in reducing the differences that existed between the various regions at the time and the less favored rural areas. Although it is explicitly stated in the Treaty of Rome (Article 39,2) that account should be taken of "structural and natural disparities between the various agricultural regions" this provision was of no concern to policy makers (Tracy, 1982). The extent to which the Common Agricultural Policy along with the other Community instruments helped rural areas in Greece will be examined in the following sections.

Early on during the period right after accession, the Greek state engaged in adjusting agricultural policy in light of the new circumstances. Policy was reoriented toward diminishing input subsidies and state intervention. The state began to play the role of the facilitator instead of the guarantor of income across the board irrespective of quantities, quality and social cost involved. Emphasis was placed on structural policies and investment aid financed by the Community and the state was channeled toward improving the efficiency of marketing and processing. Priority was given to cooperative organizations throughout the country in an effort to increase the producers' bargaining position against traders and middlemen. Attempts were made to raise the degree of self-sufficiency in certain products of the livestock sector but they were unsuccessful. The strengthening of the cooperative movement was deemed necessary due to the presence of a large number of small size family type agricultural holdings.

Several national policy guidelines were identified vis a vis the Common Agricultural Policy. One major aim was to minimize any negative impact the integration of the national sector would have on the balance of trade in European agriculture as well as to maximize any positive prospects. By means of a memorandum concerning the national economy submitted to the Commission of the European Community in March 1982, the Greek government managed to renegotiate several of the terms concerning agriculture that offered possibilities for improvement. In this context, the percentage of the Community's participation in the financing of structural programs was substantially increased. Extra, favorable provisions for Mediterranean products were pursued and the implementation of special structural programs, adjusted to the country's specific characteristics was made possible. Finally the Greek negotiating tactics proved instrumental in laying down the Integrated Mediterranean Programs.

Implementing CAP's price policy

After 1981 the management of the market for the great majority of agricultural products in Greece was gradually incorporated into the Community's Common Organization of the Markets. Domestic producers' prices for products in deficit in the Community were maintained at a certain level by adopting customs duties and variable levies as provided by the Common customs tariffs. Concerning imports into Greece the country's accession led to trade diversion enabling substantial agricultural exports (i.e. butter, barley, cheese) that would have otherwise been exported to third countries at subsidized prices to be sold to Greece at Community prices.

The Community's price policy imposed additional costs on Greek farmers as feed stuffs for livestock would cost more because of the existing support system. The profitability of domestic livestock producers was further reduced as Greek farmers ceased to enjoy the national support system which subsidized production costs. The problem faced by Greek livestock producers became acute and profit margins became very tight (Marsh, 1980).

The application of the Common Organization of the Markets in Greek agricultural production will be made more specific in the next few paragraphs (European Parliament, 1995). At the time of accession approximately 95% of the value of domestic production was covered by the eighteen existing Common Organization of the Markets. A support system was also provided for cotton according to Protocol 4 of the Treaty of Accession. The European Agricultural Guidance and Guarantee Fund (EAGGF) immediately began to replace national funding by contributing only about 9% of the gross value of production or 10% of total funding in 1982 to reach over 30% of the gross value of production or approximately 60% of total funding in 1996 (figure 4.1).

Greek agriculture has been extremely interventionist in character as compared to the rest of European agriculture. In the case of Greece the Guarantee sector of EAGGF was addressed almost exclusively to intervention practices which took up approximately 94% of total funds for the period between 1981 and 1993, whereas the respective Community average was 64%

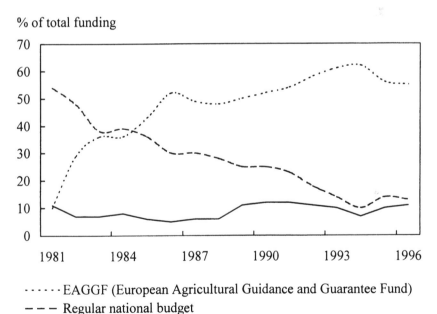

% of total funding

· · · · · · EAGGF (European Agricultural Guidance and Guarantee Fund)

− − − Regular national budget

———— National fund for public investments

Figure 4.1: **Funding by source, 1980-96**
Source: Sapounas, G., Miliakos, D., 1996

119

for the same period. EAGGF's expenditure for export restitution accounted for only 6% of total expenditure. Tobacco and cotton alone accounted for about 43% of the total cost of intervention.

More specifically, in the cereals sector, the Common Organization of the Markets initially provided for export restitutions, intervention price, the implementation of a co-responsibility levy in the case of exceeding the maximum guaranteed quantities as well as aids to small producers per unit of cultivated area. In 1988 the mechanism of stabilizers that instituted a maximum guaranteed quantity, reduced intervention prices each time annual production exceeded this level. Later on, in 1992 under the MacSharry reform, intervention prices were substantially reduced over a period of three years until 1995/1996. Compensation for price cuts was paid however on a per hectare basis adjusted according to acreage yields which prevailed in each region during a reference period in the past. This compensation is available in Greece unconditionally under a simplified system for small farmers producing less than 92 tons of cereals. Greece is a major producer of durum wheat, a product that is mainly produced in Southern Europe. Besides the general support arrangements for cereals, Greek producers of durum wheat receive a direct income aid per hectare as support to traditional production areas in disadvantaged rural regions (Tracy, 1997).

In the fruit and vegetables sector, the Common Organization of the Markets provides flexible support by authorizing producer groups to withdraw produce from the market upon compensation when prices fall below a certain level. Greece has made extensive use of the withdrawal mechanism for products like peaches, oranges, tomatoes etc. Producer groups have been considered competitive to cooperative organizations and the Greek state has only authorized the latter to manage the withdrawal mechanism. Imports are subject to tariffs, quotas and countervailing charges.

Aid is provided to the processors for certain processed fruit and vegetables of importance to Greece such as dried raisins, figs, peaches and tomatoes. Under the MacSharry reform plan withdrawal prices for fresh produce and processing aids for processed products were subject to a system of maximum guaranteed quantities and prices were reduced during periods of excess production. This scheme has already been reformed by the Council.

In the olive oil sector, price and income support policy is ruled by the same philosophy as policy implemented in the case of durum wheat. The main aim is to ensure adequate incomes for farmers in the less developed regions of the Community. Consumer prices have been kept at low levels by means of a consumption aid paid to the oil mills. A mixed system of intervention prices and producers' aid along with compensation payments to small producers whose production does not exceed 500 kgms. has been put into effect. The stabilizer system has a proportionate effect on intervention price as well as on production aid when quantities produced exceed the maximum guaranteed quantities. Small producers are exempted. The common organization of the markets provides for export restitutions and imports are subject to a modest import duty. The scheme for olive oil is about to be reformed as attempts are being made by the Commission to concentrate compensation on maximum guarantee quantities allocated to member states.

The support system for cotton provides for the fixing of an objective price and a minimum price for the product in its crude (unginned) form. A direct production aid is granted which compensates for the difference between the objective price and the actual price that prevails in the international markets. The aid is administered by the ginning enterprises that pay the producers the minimum price. If the production of unginned cotton exceeds the maximum guaranteed quantity the objective price and the production aid are proportionately decreased up to a certain limit. Quantities exceeding the limit are transferred to the next marketing year. A direct aid per unit of cultivated area is granted to small producers who cultivate less than 2.5 hectares of land. The entire process is supervised and monitored by the National Cotton Organization.

The market organization for wine incorporates intervention arrangements for excess production of low quality wine by means of distillation procedures. Community wine is protected against third country imports by imposing countervailing charges when prices fall below certain levels. Conventional customs duties have also been in effect. Surpluses have grown at a Community wide level and intervention arrangements have been reinforced under the 1988 reform to include a system of preventive distillation at low prices. Subsidies have been responsible for extensive grubbing up of low quality vineyards that has taken place in Greece. The planting of new vineyards is prohibited irrespective of type, variety and quality.

The common organization of the tobacco market originally provided intervention prices, production aids and export restitutions for fourteen variety groups in the Community. Under the MacSharry plan the scheme was radically reformed by introducing a mechanism of production control. The numerous varieties were regrouped for the entire Community into eight groups. The reform aims at stabilizing production by means of a system of production quotas reaching a maximum of 390,600 tons for the Community. Within the limits of this maximum quantity the Council has fixed the maximum guarantee funding for each group through 1997. Greece, a major producer of tobacco in the European Union, has been allocated 126,300 tons that correspond to 30,700 tons of flue cured tobacco, 12,400 tons of light air cured, 15,700 tons of sun cured and 67,000 tons of certain traditional varieties. No production aid is paid for production beyond the quota level whereas intervention and export refunds have been abolished.

In the dairy sector, the Community has set a target price for milk, just as in the case of cereals that is implemented through intervention in butter and skim milk powder. Threshold prices, variable import levies and export refunds are also effective instruments within the Common Organization of the Markets. A production quota system was introduced in 1984, according to which national quotas are distributed to dairies that allocate quantities to individual producers. The implementation of the system has proved extremely difficult for Greece that has experienced consumption and production increases which have overrun the quota of 630,000 tons. The large deficit in fresh milk cannot be satisfied through imports as it is technically impossible for Greece to import fresh milk due to the great distance with the rest of Europe.

The national milk quota for Greece accounts for 0.6% of the Community's total of 109 million tons and EAGGF expenditures for milk produced in Greece do not exceed 0.2% of the Community's total (table 4.4). The scheme has proved to be frustrating especially for young farmers who are in need of a quota, a license to produce, in order to enter the profession. National authorities hope that reconciliation will come about soon when the scheme will be reformed again.

Beef and veal meat is in short supply in Greece and is expected to remain so, thus, intervention and export restitutions have not been applied. Specific premia are granted for young male animals and suckling cows that have recently increased.

For sheep meat a basic price, uniformly applied throughout the Community determines the level of premia paid per ewe. Greece remains a major producer as well as consumer of sheep and goat meat. In 1988 the principle of a maximum guaranteed quantity had been introduced in the context of MacSharry's stabilizer's arrangements and limits exist on the number of eligible ewes per flock.

During the entire period after accession institutional prices in national currency (drachma) as provided by the Common Organization of the Markets, have steadily increased. Due to agrimonetary provisions and more specifically, due to the operation of the Monetary Compensatory Amounts (MCAs) price increases in drachmas have exceeded the inflation rate many times yielding real price increases. Negative MCAs however, have practically eliminated any positive effects the steady devaluation of the drachma would have on intracommunity trade as well as on Greek exports to third countries. Due to the difference between the green rate and the official exchange rate, in favor of the former, a situation that prevailed until 1988, MCAs had a negative effect on the country's balance of trade in agricultural products. Since 1993

Table 4.4

Milk production statistics for EU member states. The relative position of Greece (1992)

Country	Production in '000 tons	Quota in '000 tons	Quota per capita in kg	Consumption of fresh milk per capita in kg	EAGGF expenditure in ECU's	EAGGF expenditure (% of total)	National quota (% of total)
Belgium	3,514	3,310	334	107	266	6.6	3.0
Denmark	4,605	4,455	874	129	337	9.4	4.1
Germany	27,991	27,065	351	59	504	12.6	25.5
Greece	731	630	59	45	8	0.2	0.6
Spain	7,000	5,561	143	77	130	3.2	5.1
France	25,316	24,236	430	69	974	24.3	22.2
Ireland	5,382	5,246	1,499	204	169	4.2	4.8
Italy	11,068	9,930	173	56	215	5.4	9.1
Luxembourg	268	269	673	262	1	0.02	0.2
The Netherlands	10,969	11,075	743	70	1,051	26.2	10.1
Portugal	1,652	1,872	182	70	12	0.3	1.7
United Kingdom	14,783	14,590	254	125	302	7.5	13.4
Total	112,998	109,046	317*	78*	4,007	100.0	100.0

* average

Source: Ministry of Agriculture, Division of livestock production, Department of milk and milk production, Athens, April 1994 (in Greek)

the MCAs have ceased to exist. Any drachma devaluation is automatically followed by an adjustment of the green rate at any point in time during the marketing year. All the market organization measures, as described above, have had an important impact on rural development and have decisively affected the well being of the different rural regions of the country which are very diverse. They have also affected regional disparity and income distribution. Such aspects will be examined and analyzed in detail in the next section.

Greek agriculture has generally done relatively well if one judges the degree of support provided by the guarantee section of the EAGGF. Products that have historically been favored by the degree of financing are cotton, olive oil, tobacco, cereals and fruit and vegetables. The inflow of funds as a result of market policy for the period between 1990 and 1994 is presented in table 4.5. Transfers to Greece are relatively high and per capita direct payments and aid are highest for Greece than for any other member state as opposed to net trade transfers for which Greece suffers a clear net loss (figure 4.2).

These is no evidence that funds have been constructively utilized by farm holdings and cooperatives in particular, which in many cases have administered the implementation of the Common Organization of the Markets. Financial support of this nature, however has undoubtedly prevented a serious income deterioration.

Implementing the common socio-structural policy

At the outset of Greece's integration into the European Economic Community, CAP's structural policy was seen as a means for structural improvement. In practical terms, the extent and effectiveness of the 1972 directives on farm modernization, retirement and retraining that introduced the common socio-structural policy (Mansholt Plan) have had no effect as they were not aimed at dealing with the problems of small and disadvantaged farms.

More specifically, by means of Directive 72/159, concerned with farm modernization, aid in the farm of an interest rate subsidy or capita grant on investments was granted to farmers who submitted a development plan showing that, upon completion, the level of earned income would compare to that received for non-agricultural work in the region. The scheme was out of reach for practically all Greek farmers. Directive 72/160 encouraged elderly farmers to retire and release their land to the benefit of more competitive farms whereas Directive 72/161 was concerned with providing guidance for the acquisition of agricultural skills.

Greek farmers did not embark on development plans as they encountered several obstacles such as high interest rates as compared to incentives offered, high unemployment rates in the non-agricultural sectors, difficulties in attaining the comparable income criterion which practically excluded the great majority of farms and the lack of farm accounting records.

Greek farmers took advantage of Directive 75/268 that introduced compensatory allowances in mountainous and other less favored areas with an aim to maintain population in such areas. The directive was applied to approximately 18% of the total number of farms in Greece which, in 1990, accounted for about 8% of total financing for compensatory allowances in the Union.

Table 4.5
EAGGF (Guarantee) transfers to member states for market support (1990 - 1994)

	1990	1991	1992	1993	Million ECUs 1994
B	873.7	1,468.5	1,378.2	1,298.7	1,170.4
DK	1,113.7	1,220.3	1,166.8	1,334.7	1,278.4
D	4,355.2	5,234.5	4,830.5	4,976.2	5,179.9
GR	1,849.7	2,211.2	2,231.4	2,715.0	2,718.9
E	2,120.8	3,314.3	3,578.1	4,175.7	4,408.3
F	5,142.2	6,394.4	6,916.5	8,184.8	8,001.2
IRL	1,668.4	1,731.1	1,452.8	1,649.9	1,480.0
I	4,150.3	5,353.4	5,141.5	4,765.4	3,460.6
L	5.2	2.8	1.1	7.3	12.1
NL	2,868.7	2,679.3	2,389.8	2,328.1	1,916.0
P	214.2	315.6	423.8	478.1	708.4
UK	1,975.9	2,391.3	2,451.1	2,737.9	2,939.0
Community [a]	215.5	69.2	145.9	96.4	139.0
EU12	26,453.5	32,385.9	32,107.5	34,748.2	33,412.2

[a] Direct payments from the EU to beneficiaries

Source: *First report on economic and social cohesion 1996*, European Commission Luxembourg 1997

ECUs per capita

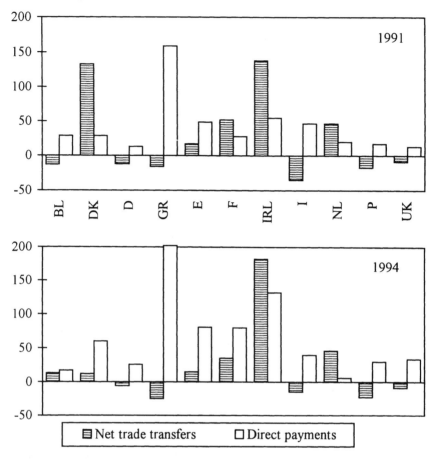

Figure 4.2: **Flows as a result of the CAP-market Policies**
Source: *First report on economic and social cohesion 1996,* European
Commission Luxembourg 1997

Extensive use was made in Greece of policies for the improvement of
structures in marketing and processing (Regulation 355/77) on the basis of
national programs for a number of sub-sectors. Substantial interest for grants
has been expressed for the construction, expansion, relocation and
modernization of processing plants in the cereals, wine, olive oil, fruit and
vegetables, tobacco and the livestock sector. Requests for grants by
cooperatives and the private sector have always been much higher than the
funds available.

The overall limited application of the structural measures in the Community's south, led to an awareness of practical problems the Mediterranean regions of the Community were faced with. A Mediterranean package of measures was designed and a program was implemented, in the case of Greece (Regulation 1975/82) for the acceleration of agricultural development in certain less favored areas extending from Epirus in the north-west to the southern part of the Peloponnese. Similar programs were implemented in the Mezzogiorno region of Italy and the Mediterranean regions of France. Such Community awareness for a structural policy tailored to the needs and characteristics of specific regions became more apparent when Greece pressured the European Council at the level of heads of state and prime ministers for a shift in emphasis within the Common structural policy away from structural adjustments of the "Mansholt type". They pushed for a policy that could contribute to rural development in peripheral areas of the Mediterranean, having in mind the prospect of accession of Spain and Portugal in 1986. The ambitious "Integrated Mediterranean Programs" were introduced in 1985 to cope with socioeconomic problems of the Community of the ten member states and to face up to competition from the two prospective new members and more specifically, Spain. Later on in 1988, when the structural funds reform was agreed, this integrated, region specific approach was carried even further.

In 1985 the basic sociostructural directives were replaced by Regulation 797/85, later amended by Regulation 2328/91, which allowed for the provision of investment aids to practically any lower income farmer. The prohibitive lower comparable income criterion for eligibility was replaced by an upper limit according to which the improvement plan submitted by the farmer should not lead to an income increase over 120% of a reference income. A stronger link with market policy was enforced, quality production was encouraged according to market requirements and agricultural practices friendly to the environment were promoted.

For the first time this policy framework provided Greek agricultural holdings with the opportunity to carry out investments in improvements such as land reclamation works, the construction of agricultural buildings, the purchase of machinery and live animals, and general infrastructure improvements. Investments envisaged the installation of young farmers but were proved relatively ineffective in Greece, due to constraints related to the availability of land appropriate for cultivation and low compensation rates.

A major shift in policy toward the integrated, region specific approach took place in 1988 when the Community of the twelve member states doubled spending on structural action. The structural funds, namely the European Regional Development Fund, the European Social Fund and the Guidance Section of EAGGF were reformed to serve specific priority objectives.

Development and structural adjustment of regions whose development was lagging were served by Objective 1. The entire territory of Greece, Portugal and Ireland, a great part of Spain, the southern part of Italy-Sardinia, Corsica and French territories where the per capita Gross Domestic Product (GDP) was less than 75% of the Community's average was defined as priority. Objective 1 funds were defined as Objective 5b regions on the basis of income, the level of socioeconomic development and the share of agriculture in total employment. Such regions were allocated funds for similar development activities even though the level of funding was not as extensive.

Agricultural measures of a structural nature provided by legislation existing before the reform of the structural funds remained in application throughout the Community with exclusive EAGGF funding. Horizontal type actions, as opposed to regionalized measures, aiming at speeding up the adjustment of agricultural structures with a view to the CAP reform were carried out in areas designated as Objective 5a regions.

Rural development measures, social assistance to farmers, improvements in marketing and processing, improvements on farms, aid to young farmers, early retirement schemes and measures appropriate for adjusting production were implemented horizontally as well as integrated into regionalized programs, in the context of the First Community Support Framework (1989-1993). Community assistance accounted for approximately 75% of the total cost of such measures applied in Greece. The incorporation of structural measures in agriculture into the Operational Programs of the First Community Support Framework constituted the beginning of a new Community policy that attempted to resolve farm problems in an integrated manner and not in isolation without taking into account the general economic environment.

Integrated rural development

The Maastricht Treaty of 1992 reinforced the conviction for an economic and social cohesion throughout the Union. The Treaty provided for the establishment of a new Cohesion Fund that would make additional funding available in order to serve the purpose of reducing disparities among the various levels of development of the different regions, including rural ones. Spending on the structural funds was also increased over the five year period between 1994 and 1997. The new structural policy is characterized by a clear integrated rural development dimension. It has territorial focus and takes into account all sectors of the economy.

In Greece, government action gave emphasis to providing infrastructure in rural areas tailored to the needs and requirements of the country's regions. Thirteen regional operational programs were prepared corresponding to the country's thirteen administrative districts. Several sectoral programs were also incorporated into the Second Community Support Framework. The Ministry of Agriculture had the responsibility for designing and implementing one such program. Each regional operational program was characterized by specific problems and opportunities. Agriculture is included in the development effort and it plays an important role in all rural areas in the country. Agricultural actions are geared toward different objectives according to the regional context. Measures are funded by all three structural funds, the new Cohesion Fund and the European Investment Bank. Community initiatives such as the LEADER program promotes integrated rural development in all 50 prefectures of the country whereas the INTERREG Program provides assistance for infrastructure improvements at border regions.

Total EU funding for structural support during the five year period is estimated to approach 3 billion ECUs per annum, bringing up total spending including national funds to an annual average of approximately 5.7 billion ECUs. These figures correspond respectively, to 3.7% and 7.2% of the National Gross Domestic Product (table 4.6). According to estimates, total structural support is expected to contribute to the increase of the GDP by 2.5% (Maraveyias, 1992). Financial contribution for structural support in Greece by the European Structural Funds was increased between the First and the Second Community Support Framework by over 50%.

Greece's share in total EU structural support distributed among the Member States has deteriorated from 12.5% to 10.6% between the five year period of 1989-1993 to 1994-1999 whereas Spain and Germany increased their shares from 20.6% to 25.3% and 11.5% to 13% respectively (figure 4.3). Similarly, annual per capita support for Greece has increased between the two periods from approximately 200 to 300 ECUs. Nevertheless, in 1991 Greece had the lowest per capita GDP in purchasing power equivalent units, having recently been outflanked by Portugal which in 1986 demonstrated the lowest per capita GDP in purchasing power equivalent units (figure 4.4).

The sectoral operational program for agriculture implemented in Greece along with the regional programs provided funds to support actions aimed at improving the competitiveness of agricultural activity as well as safeguarding the social fabric of the countryside. In this context, the EAGGF financed the implementation of Regulations 2328/91 and 866/90 that amended Regulation 355/77. Funds directly addressed to agriculture did not exceed 15% of the total, a percentage that indicated a substantial drop as compared to the respective figure for the period between 1989 and 1993.

The regional and sectoral programs implemented in Greece, stemming from the Community's Support Framework did not constitute a coherent national rural development policy. The various Community instruments had demonstrated various degrees of effectiveness and did not act in a comprehensive and integrated manner. Certainly some of the available schemes have to be further adjusted in order to respond effectively to the people's needs in the regions. It is also certain that the effectiveness of all supportive mechanisms at the national level, namely, public institutions and organizations have to be substantially improved in order to cope with emerging new challenges.

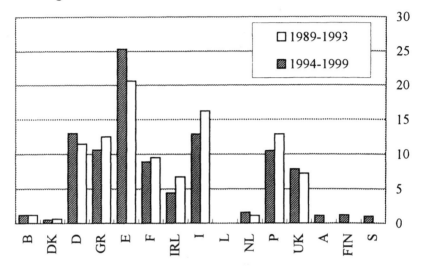

Figure 4.3: **Structural support. Distribution of finance among member states (%), 1989-93, 1994-99**

Source: *First report on economic and social cohesion 1996,* European Commission Luxembourg 1997

Environmental measures

Historically, the Greek state was reluctant in applying environmental measures in agriculture as provided for by the first Common Agricultural Policy reform. Measures provided either in piecemeal form by various regulations just as 797/85, or the different set aside schemes, or a more consolidated and comprehensive form such as the latest agri-environmental schemes were not widely applied. The environmental effects of modern farming methods have not, as yet, generated increasing public concern as they have in several of the northern European Union member states. No stringent environmental controls were introduced that would alleviate water and air pollution from the environment in spite of a host of measures that were introduced at the Community level in recent years. Greece still has not responded properly.

Table 4.6
Total structural support (including the cohesion fund and community initiatives) 1989-93, 1994-99

| | 1989 - 1993 (annual average) | | | | 1994 - 1999 (annual average) | | | |
| | Million ECU's | | % of GDP (aver. 89-94) | | Million ECU's | | % of GDP (1994) | |
	total spending	EU contrib.	Total	EU	total spending	EU contrib.	Total	EU
B	485	173	0.30	0.11	1,089	349	0.57	0.18
DK	274	86	0.26	0.08	426	140	0.34	0.11
D	6,741	1,680	0.53	0.13	13,954	622	0.81	0.21
GR	3,091	1,834	4.47	2.06	793	2,956	7.20	3.67
E	6,201	3,017	1.54	0.75	13,747	7,066	3.38	1.67
F	4,114	1,387	0.42	0.14	7,107	2,491	0.63	1.74
IRL	2,212	980	5.99	2.66	2,180	1,234	4.98	0.22
I	5,485	2,374	0.63	0.27	9,722	3,608	1.13	2.82
L	41	15	0.45	0.17	57	17	0.49	0.42
NL	488	163	0.21	0.07	1,498	436	0.53	0.15
P	3,789	1,892	6.15	3.07	5,300	2,940	7.17	3.98
UK	2,659	1,066	0.34	0.13	4,779	2,164	0.56	0.25
EU12	35,580	14,666	0.71		65,651	27,024	1.11	0.45
A					1,572	316	0.94	0.19
FIN					1,134	331	1.38	0.40
S					878	261	0.53	0.37
EU15					69,235	27,932	1.12	0.51

Source: *First report on economic and social cohesion 1996*, Luxembourg European Commission 1997

131

The Single European Act of 1986 provided the legal basis for the design and implementation of agri-environmental measures. The Nitrate Directive 91/676 aims at reducing water pollution caused by nitrates from agricultural sources. It required that member states establish a code of good agricultural practice to be implemented by farmers on a voluntary basis. Vulnerable zones were designated in order to implement action programs in the context of which certain measures become mandatory.

In addition Regulation 2092/91 on organic farming specifies cultivation practices to be followed in order for a product to be classified and labeled as "organic". In the context of the common structural policy, initiatives have been taken to encourage environment friendly farming practices. The protection and improvement of the environment have been incorporated as a criterion of eligibility when granting aid for investments on farms. Special aids were made available to farmers, in the context of Regulations 797/85 and 2328/91, that agree to comply to production practices which conform with requirements for protecting the environment and landscape in Environmentally Sensitive Areas (ESAs). Finally, the MacSharry reform of 1992 provides for an environmental scheme under which aid is granted to farmers who adopt practices which reduce pollution or agree to set farmland aside for environmental purposes (Regulation 2078/92). Greek agriculture has not been affected by incentives provided in the context of the Community's agri-environmental policies.

In the area of biological agriculture, certain progress is currently being observed in Greece. Organic farming is a suitable operation for the country's agricultural sector given the structure is dominated by small family farms. A number of farm managers who operate biological farms have been certified since 1993 but the development of biological agriculture in Greece still remains at relatively low levels as it does not exceed 0.2% of the total cultivated agricultural land. Conversion from conventional to biological agriculture has been viewed by Greek farmers as a difficult procedure with risks from both the technical and the economic point of view as it implies higher costs and lower yields. The necessary information on all aspects of organic farming has not been made readily available to Greek farmers. A well developed distribution network for these products is lacking in spite of an increasing market demand for organic products. Another problem that farmers face is related to the situation concerning land rentals. According to the European Union's standards for organic farming, a product can be certified as organic three years after ceasing the use of agrichemicals. Many farmers in Greece are not able to secure a long term rental contract. Biological agriculture, along with the production of specialized and location specific products may contribute to the solution of income problems, especially in mountainous, coastal, island and less developed regions in the country. In such regions where extensive farming systems prevail, conditions are quite favorable for converting from conventional to biological cultivations. In such rural areas, climatic, social and structural characteristics are favorable.

The Nitrate Directive that sets limits on the net delivery of nitrogen to the soil and water has not been applied in Greece, in spite of the acute problem that exists with surface and subsurface water resources especially in certain regions. The code for correct agricultural practice was issued by the Ministry of Agriculture in 1994 but it has not been institutionalized yet. The three accompanying measures of the 1992 MacSharry reform are currently being applied, namely Regulations 2078/92 (agri-environmental measures), 2079/92 (early retirement) and 2080/92 (afforestation). National policy concentrates on the management of mountainous and island regions which are under pressure for abandonment, the protection against soil erosion, the preservation of areas of specific ecological significance and the protection of flora and fauna, the protection of the biogenetic material, the protection of natural resources (especially water) against pollution and the protection and promotion of biodiversity and the rural landscape. Currently three sub-programs are under way in the context of Regulation 2078/92 concerning: 1. biological agriculture that is applied in the entire country covering 1500 farms, 2. the reduction of nitrate pollution in certain zones which refer to 750 farms only, and 3. the long term set aside. Two additional sub-programs are about to be put into effect: 1. a program for the protection and the promotion of biodiversity and genetic diversity of wildlife habitats and species threatened by extinction and genetic erosion respectively and 2. a program for the preservation of areas of special ecological significance by reducing pollution agriculture causes to the environment as well as by extensifying livestock production. Three more sub-programs have been prepared and await approval before implementation; one for the protection against erosion, one for the preservation of unusual and rare landscapes and one for the promotion of farmer training and education. Finally, certain countryside measures are also part of about fifty local initiative projects in the framework of LEADER II.

Assessment

It has been some time since the Common Agricultural Policy drifted away from price supports toward direct income subsidies. This development came about at a time when it became very clear that the effect price supports have had on farm incomes, income distribution and rural development was rather ineffective in spite of the very heavy burden it placed on the Community's budget. On the other hand, the 1992 radical policy reform was more or less unavoidable as the CAP would have to reorient itself in order to conform to the General Agreements on Tariffs and Trade (GATT) guidelines as explicitly

Annual per capita support in ECUs

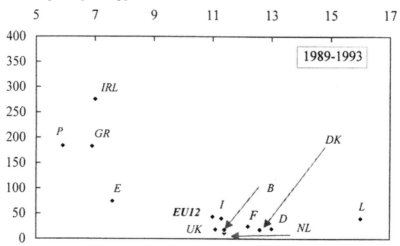

Per capita GDP in '000 of purchasing power equivalent units 1986

Annual per capita support in ECUs

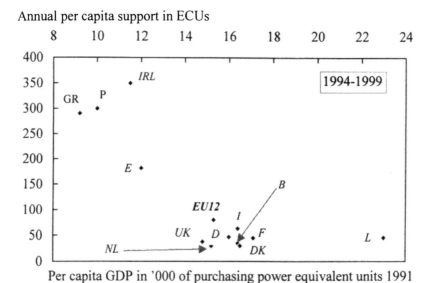

Per capita GDP in '000 of purchasing power equivalent units 1991

Figure 4.4: Structural support as compared to GDP by member states

Source: First report on economic and social cohesion 1996, European Commission Luxembourg 1997

stated in the Punta del' Este Uruguay Round Agreement on Agriculture. Furthermore, it is certain that under World Trade Organization rules, price supports and subsidies in general that are attached to product quantity will have to be further dismantled as there will be a tendency for an across the board elimination of domestic support.

A substantial restructuring of Greek agriculture aiming to increase productivity, enhance product quality and reinforce international competitiveness in order to reduce its international trade deficit will be necessary. Agriculture in Greece had been receiving substantial support before 1981 from national funds. Such support was further strengthened and augmented in the context of the Common Agricultural Policy during the period between 1981, the year of accession and 1997 as agricultural activity was supported by both national and Community funds.

Since 1981 national funds contribute to the implementation of all sociostructural measures as provided by EU common policy such as to compensate allowances in mountainous and less favored areas, early retirement schemes, all investments concerning the improvement of the conditions in marketing and processing as well as the farm modernization schemes. Such participation in funding by national sources has varied between 30% and 50% of total contribution for sociostructural instruments (Sapounas, Miliakos, 1996). The role of national financial support is best realized when account is taken of expenses related to interest subsidies for mortgage loans for rural citizens, administration expenses for the operation of the costly system that manages the Common Organization of the Markets in Greece. In addition, national expenses for social security, protection against natural disasters and pensions for farmers are very high.

The extent to which the various segments of the Greek farm population have benefited from the operation of the Common Organization of the Markets and generally from national and Community agricultural policy measures is debatable. It is very likely that such policies have protected incomes in the short run and have encouraged some rural households to stay in farming on an exclusive basis. The great majority of Greek farmers has moved to other more profitable activities whereas a third category, probably the younger farmers have already abandoned the countryside. Price support has certainly benefited the larger producers who are very vocal in demonstrating against price support increases. Structural policies pursued seem to have been more instrumental in managing what they were designed to do; keeping people on agricultural land and the countryside as Greece still maintains a relatively large number of small farms.

Replacing price supports with direct payments to farmers, intended to compensate for income losses on a transitional basis due to the CAP reform, although complicated and expensive to administer, has proved to be a better agricultural policy instrument on economic and equity grounds.

One should expect that in Greece, just as elsewhere in the Community, price falls coupled with the imposition of quotas in production should normally lead to a supply reduction. Price falls will certainly be responsible for a shrinking of gross margins which will prove to be critical for agricultural holdings with a high cost per unit of production. On the other hand, consumption especially of products that demonstrate a high price elasticity should be expected to increase. Such developments will certainly have a clear impact on the balance of trade of agricultural products. Of course, the expected decrease of consumers' prices, due to cheaper imports of products for which Greece is deficient such as dairy and beef meat, will have a favorable impact on consumers' surplus.

Agricultural policy changes as a consequence of the CAP reform and the Uruguay Round Agreement will make it necessary for Greek farmers to acquire competitive skills in an environment which will be increasingly dominated by the operation of free market rules as intervention practices and price supports will be gradually phased out.

Greek agriculture will have to prove more competitive at the level of production, processing and marketing. Structural adjustment, the enhancement of human capital, the improvement of infrastructure, concern about the quality of the environment along with social protection provided by the state with farmers' contributions should be the main priority guidelines of national policy.

Bibliography

European Parliament (1995), *Incidences Financieres du Fonds Europeen d' Orientation et de Garantie Agricole en Grece, de l'Adhesion (1981) a la Creation du Marche Unique (1992),* Serie: Agriculture-Peche-Forets, W-26/I, Direction Generale les Etudes: Luxembourg.

Maraveyias, N. (1992), *The Process of European Integration and Greek Agriculture in the 90's,* Papazisis: Athens. (in Greek)

Marsh, J. S. (1980), 'The CAP and Greece. A Community Perspective', *Food Policy,* Vol. 5, No. 4.

Pepelasis, A. (1980), 'The CAP and Greece. The Implications for Greek Agriculture', *Food Policy,* Vol.5, No. 4.

Sapounas G. and Miliakos, D. (1996), *Greek Agriculture in the Post-War Era. Impacts from Accession into the European Union and lessons for the Future Agricultural Economics,* Studies Series, No. 51, Agricultural Bank of Greece: Athens. (in Greek)

Tracy, M. (1982), *People and Policies in Rural development. Institutional Problems in the Formulation and Implementation of Rural Development Policies in the European Community*, The Arkleton Trust.

Tracy, M. (1997), *Agricultural Policy in the European Union and Other Market Economies*, 2nd edition, Agricultural Policy Studies: Genappe, Belgium.

5 The impact of policies for the rural sector

The evolution of agricultural production and trade

Since 1981 the implementation of the Common Agricultural Policy in the cereal sector has had several implications. The relative importance of cereal production with respect to the total value of agricultural production has been steadily decreasing. In absolute terms, total production of cereals has remained more or less constant in spite of the substantial reduction (10%) of the total area cultivated under cereals.

The average size of agricultural holdings specializing in cereal production has remained approximately steady, at the level of 12 hectares that corresponds to about one fourth of the Community's average. Average yields of cereal production in Greece do not exceed 75% of the Community's average whereas net returns, per labor unit, remain at approximately 50% of the Community's average.

Land is rented out to cereal farmers at rates that are 50% higher than rates observed in the Netherlands, the Member State with the highest rents second to Greece. Cereal producers' incomes in Greece are not comparable to the average incomes of cereals producers in the Community. Low producers' incomes are explained by the small size and fragmentation of holdings as well as the irrational use of the means of production (European Parliament, 1995).

The mix of cereal production in Greece has changed dramatically since 1981. The production of common wheat has decreased by approximately 60% while the production of durum (hard) wheat has doubled. Price differentials as well as the application of direct income aid per unit of cultivated area to durum wheat producers have been responsible for such a development. In

139

addition the introduction of hybrid corn seeds has led to a spectacular increase of corn yields which exceed the Community's average by approximately 25%.

The application of the Common Agricultural Policy and the high prices paid to cereal producers after the accession along with the adoption of the Common Customs Tariffs have had a positive impact on the country's international trade of cereals. Greece, from a net importer of cereals, in 1981 (1.8 billion drachmas), became a net exporter in the beginning of the 1990s (28.6 billion drachmas in 1991).

Real producer price increases, in national currency were mainly due to the devaluation of the green drachma. However, producers' incomes remained more or less stable throughout the period of reference as input subsidies were gradually eliminated. Consequently, the cereals sector, with the exception of high quality corn, does not demonstrate any signs of productivity and quality increases that would guarantee increased competitiveness versus the rest of the Community.

On the contrary, the sector remains very much dependent on intervention practices and export restitutions. The situation becomes difficult if one takes into account the diminishing role such policy schemes will play in the future due to policy reform and international obligations. Further price decreases will be detrimental to the farmer as average cost per unit of production remains very high.

In 1981, fruit and vegetables accounted for approximately 27% of the total value of agricultural production. In the early 1990s the sector's relative importance was somewhat reduced to slightly under 25%. Agricultural policy schemes for fruit and vegetables under the Common Organization of the Markets provided a rather limited degree of protection and support. The Community's funds devoted to the implementation of agricultural policy instruments were inferior compared to other sectors and were highly concentrated on measures of an extreme interventionist character such as withdrawing quantities from the market and subsidizing exports. The type of measures implemented, as well as the limited level of support was a result of the Community's political conviction to offer concessions to third countries by encouraging imports from Mediterranean non-member states.

Producer price increases observed during the early years after accession were followed by production increases. Production was later stabilized at lower levels as cultivated areas under fruit and vegetables, with the exception of citrus fruit, were decreased and yields deteriorated due to weather conditions that prevailed for a number of years. The production of fruit and vegetables in Greece, as in the case of cereals, has faced serious problems, mainly due to the high cost of transportation to various markets in the Community and the perishable character of the fresh produce.

In addition, high production cost per unit of cultivated area is related to the small size of the average holding in Greece. In the case of fruit, about 5 hectares is an average size whereas in the case of vegetables, it is less than 3 hectares. Producers' income for fruit is lower than 75% and 30% of the average income in the Community for each one of the two product categories respectively (European Parliament, 1995).

Fruit and vegetables constitute the most important sector as far as exports are concerned. In 1981, exports of fruit and vegetables from Greece represented about 10% of total exports by value, or 37% of the total value of agricultural products exported. During the reference period, since 1981, fruit and vegetables' importance in the country's international trade increased slightly.

Although Greece remains one of the major exporters of fruit and vegetables in the Community after Spain, Italy and France, agricultural policy measures enforced throughout the same period, did not prove instrumental in enhancing the country's competitive position. Greece maintains a leading position as an exporter of certain fresh and processed fruit and vegetables such as peaches, citrus fruit, apricots, tomato paste, raisins and cucumbers. However, no substantial progress has taken place in the areas of vertical integration of production, processing as well as reorienting production in order to take advantage of important markets in the Community.

In spite of the important positive balance in international trade of fruit and vegetables, the way agricultural policy was implemented in the sector did not lead to substantial production increases in off season fruit and vegetables. It did not bring about substantial improvements in quality, packaging and processing.

On the contrary, a good part of marketable production was directed to intervention (withdrawal). In many cases intervention prices coincided with the market prices for a number of products. Withdrawal for apple, peaches, citrus fruit, etc. has acted as a counter incentive against the restructuring of production that is very much needed in order to enhance the country's competitive position (European Parliament, 1995).

The Common Organization of the Markets for olive oil takes into account the important fact that olive oil is produced in the less favored areas of the Community, mainly by numerous small producers. Favorable Community policy elements have provided Greek producers with a minimum intervention price along with production and consumption aids.

Agricultural policy implemented since the year of accession has led to important production increases. In 1981, as well as throughout the reference period, olive oil production in Greece accounted for a little less than 8% of the total value of domestic agricultural production. However, the total volume of production, in absolute terms has increased from approximately 250,000 tons in 1981 to an annual production of over 300,000 tons in the 1990s. Total cultivated area under olive trees has increased from about 600,000 hectares in 1981 to over 700,000 hectares in the 1990s - an area that corresponds to a number of olive trees that exceeds 130 million (European Parliament, 1995). Average yields are low as compared to Spain and Italy (roughly about 2.5 kilograms per tree) but olive oil quality is superior to that produced elsewhere in the Community.

Producers' prices as well as production aid have steadily been increasing in real terms since accession. Exports have followed an upward trend reaching an annual average of about 100,000 tons in the 1990s. Exported quantities have increasingly been directed to the Community's markets and more specifically to the Italian market. Greece has not been helped by policies implemented to increase the value added for its exports of high quality olive oil by shifting from bulk exports to exports in small containers that could be identified for their contents' quality and place of origin.

Greece accounts for about 80% of the total amount of cotton produced in the Community. The rest of the cotton production is basically accounted for by Spain. In 1981 unginned cotton production in Greece corresponded to only about 4.5% of the total value of agricultural production and occupied 4.5% of the total cultivated area in the country. The generous Community policy scheme implemented throughout the period since the country's accession has created favorable economic conditions and provided incentives for the expansion of production. By the mid 1990s cotton production increased in a spectacular way and currently accounts for over 13% of the total value of agricultural production.

Cultivated areas under cotton production have also expanded in a similar manner taking up 40% of the total irrigated area of the country. Substantial quantities of competitive products at the level of production such as sugar beets, wheat, corn and tomatoes for industrial use have been replaced with cotton in irrigated farm land. Cotton production in the country has reached the maximum quantities guaranteed by the Common Organization of the Markets and is currently at a level over three times higher than that registered in 1981.

Final producers' prices in national currency have steadily increased but have now reached a maximum level as further increases of production implies the activation of a stabilizer mechanism that imposes levies or actual cuts. Production is concentrated in the irrigated plains of Thessaly in Central Greece, Macedonia and Thrace in the north and the north-east as well as in the region of Central Greece, north and north west of Athens.

The way policy schemes have been applied in Greece has led to a deterioration of the product's quality for a good part of total production in spite of the fact that certain domestic cotton varieties have been known in the past for their excellent quality. Cotton production has also caused excessive misuse of water resources. The excessive use of fertilizers and other chemicals have had a serious negative impact on the environment and natural resources. Details on cotton's participation in the balance of trade of agricultural products are presented in future chapters (European Parliament).

Common policy instruments for Greek wine were designed to discourage supply. Production as well as cultivated areas have been reduced substantially since the country's accession in the European Community. Since 1981, production has been reduced by over 13% and cultivated areas have decreased by almost 35%. This development is explained by wine producers' willingness to engage in structural   programs partially financed by the Community that provided the incentive for the uprooting of low yield, marginal vineyards which produced low quality table wine throughout the country. The negative economic environment has manifested itself by the low common institutional prices that have prevailed.

In general, wine production in Greece has not been in a position to cope with intra Community competition in spite of the existence of a great variety of good quality wines. The average vineyards in Greece are smaller than half the average size of the Community's vineyards. Producers' incomes, in money terms, are about half the incomes in the rest of the Community. In addition, wine production in Greece is negatively affected by the low per capita consumption of wine in the domestic market that is currently less than half of the per capita consumption in the other wine producing countries in the Union.

Recently, however, high quality wines have been produced in small quantities by newly established wine factories that operate on a small scale. Per capita consumption of high quality domestic wine has also been steadily increasing in Greece. Imports of wine into the country have demonstrated an upward trend since 1981 whereas exports have never been substantial as the Greek product has not been able, as of yet, to make a dent in the foreign markets of Europe and elsewhere (European Parliament, 1995).

Tobacco has traditionally been a product of great significance for Greece. Its relative importance has decreased slightly but it still remains a product of exclusive importance for certain less fertile marginal areas in the regions of Macedonia, Thrace, Thessaly and the prefecture of Etolia-Akarnania in Western Greece where the production of practically any other product cannot be maintained on purely economic grounds.

Tobacco still accounts for about 6% to 7% of the total value of agricultural production. Greece has been and still is the most important producer and exporter of tobacco in the Community.   Approximately   half of the Community's tobacco is still produced in Greece. Immediately after the reform of the Common Organization of the Markets the total area cultivated under tobacco in Greece was substantially decreased due to the implementation of the quota system and the elimination of intervention and export restitutions.

Greece has experienced a substantial restructuring of tobacco production that was induced by the common policy implemented. Greek farmers proved extremely capable of adopting new varieties. Production shifted away from certain traditional sun cured varieties of the oriental type to light air cured and flue cured varieties such as Virginia tobacco which now takes up about 40% of total tobacco production in the country.

Although the value of tobacco exports to the rest of the Community did increase during the second half of the 1980s, the value of tobacco exports as a percentage of the total value of agricultural exports decreased from 18% in 1981 to 13% in the first half of the 1990s (European Parliament, 1995).

Livestock production in Greece never developed to proportionate levels with those prevailing in most member states of the Union. The small size of livestock production units, coupled with inappropriate pasture land and insufficient production technologies are the main reasons that explain the high production costs, low yields and low economic returns, which do not compare with the Community's average. The adoption and implementation of the Common Agricultural Policy measures have not alleviated the sector from its structural stagnation problems.

In 1981 livestock production, in total, represented approximately 30% of the total value of agricultural production. Since the year of accession per capita consumption of meat and milk products has increased by about 30%. This increase in consumption was not followed by a similar increase in production in spite of the positive effect common institutional prices have had on domestic prices. Consequently the degree of self sufficiency for beef meat dropped to less than 28%. International trade was diverted and imports from Greece's neighboring countries in Central and Eastern Europe were substituted by high priced imports from the European Member States.

More specifically, in the case of the beef meat sector the number of animals decreased by about 25% since 1981. Apparently the positive trend manifested in producer prices was counterbalanced and offset by an increase in the cost of production associated with an increase of input prices that came about, among other reasons as a result of the elimination of input subsidies. The implementation of the Common Organization of the Markets in the beef meat sector in Greece has resulted further in a substantial reduction of the number of medium and small livestock units (with less than 40 animals) while at the same time an increase in the number of large units was observed. It should be noted that almost 90% of the sector's livestock units in Greece raise less than 9 animals each.

In the ten year period following the country's accession into the Community, the number of small and medium livestock farms was reduced to half. The number of large units increased by approximately 30%. During the same period, prices for foodstuffs more than doubled. The value of beef meat imports is a principle factor in determining the general balance of trade, next to energy imports in importance, and has tripled since 1981 (European Parliament, 1995).

In the beginning of the reference period, sheep and goat meat accounted for over 9% of the total value of production. However, the respective percentage dropped to approximately 6% by the mid 1990s. The level of self sufficiency was reduced accordingly from about 95% in 1981 to less than 85%. This development is explained by an increase in consumption as well as by the relative stagnation of production during the period after the adoption of policy measures incorporated in the Common Organization of the Markets. Throughout the same period, institutional common price increases did not exceed the inflation rate in nominal terms. The direct income aid granted by Community and national funds for each "eligible" animal in each flock as a structural measure was instrumental in compensating income losses especially in mountainous and less favored areas where most of the actual production is maintained (European Parliament, 1995).

The poultry sector in Greece is highly concentrated as a small number of capital intensive units account for a great part of total production. During the period after accession production increased by about 40% following an important increase in per capita consumption. Competition among major producing firms has resulted in a substantial reduction of prices at the retail level in spite of the high production costs and low productivity characterizing domestic production. The country's degree of self sufficiency has dropped slightly, allowing for a modest increase of imports.

The capital intensive pork meat sector is similarly characterized by a high degree of concentration at the production level. After accession the sector was confronted with fierce competition from relatively cheap imports from other Member States of the Community. Producer prices have not been favorable for the high cost domestic producers. Total domestic consumption increased by 65% between 1981 and the mid 1990s and consequently the degree of self sufficiency dropped to about 65%, the lowest level ever recorded (European Parliament, 1995).

Throughout the post accession period the production of milk and milk products has been maintained at approximately the 1981 level. Cheese production has gradually increased by almost 25% whereas the production of butter has dropped substantially. Milk production per cow has increased since 1981 by 35% but it corresponds only to 70% of the average production in the Community.

The average number of dairy cows per farm has doubled since 1981 to exceed 4 cows per farm compared to over 18 dairy cows in the Community. Due to an increase in per capita consumption of milk and milk products in Greece, the degree of self sufficiency has dropped for fresh milk, cheese and especially for butter. The value of imports has increased dramatically as a result of the stagnation of production, the increase in consumption and the higher prices paid for products of Community origin. In the case of beef meat, it was replaced by imports from neighboring third countries (European Parliament, 1995).

The degree of support in Greek agriculture during the period after accession (1988) is presented in table 5.1 by means of the Production Subsidy Equivalent-PSE index (Bazioti and Bourdaras, 1990). PSEs correspond to the value of transfers for producers as a result of the implementation of a host of agricultural policy measures such as price supports, direct payments, input subsidies, variable import levies, import duties etc. Thus, the PSE index takes into account support stemming from the market as well as any financial transfers from the national or the Community budget. Support for Greek agriculture in general was at the level of 40% of the gross value of agricultural production, adjusted for any direct income aids granted to producers.

Support for crop and livestock products was found to be at the level of 36% and 48% respectively. As expected, most livestock products are classified in the category characterized by a high degree of support (milk, beef meat, eggs, poultry). Certain crops such as tobacco, cotton and dried raisins receive high financial support as opposed to practically all the other crops produced in Greece.

The highly interventionist character of agricultural production in Greece is demonstrated by EAGGF's Guarantee section expenditures expressed as a percentage of the value of agricultural production (table 5.2). During recent years, such expenditure takes up over 30% of the total value of agricultural production.

The EAGGF's total Guarantee expenditure's share in the total value of agricultural production is highest for Ireland (34.4% in the year 1994), the only Member State that precedes Greece. However, support from public funds, expressed as a percentage of the total value of agricultural production has been following a downward trend in Ireland since the beginning of the 1990s whereas for Greece the trend is reversed.

The high share of expenditures observed in Greece is accounted for by the high cost of implementing price support, intervention and withdrawal practices as provided by the Common Organization of the Markets.

In absolute terms, however, EAGGF expenditure in Greece per holding for price and income support (Guarantee section) is below the Community's average. For the year 1993, for example, average expenditure per holding was 4300 ECUs as compared to 5500 ECUs of the Community's average. Agricultural holdings in the Netherlands, Belgium and Denmark received between 17,000 and 20,000 ECUs each on average.

Table 5.1
Producer subsidy equivalents for selected agricultural products in Greece (1988)

Product	% in total value of agricultural production	Producer subsidy equivalent
Common wheat	2.3	35
Durum wheat	2.7	38
Maize	4.5	47
Olive oil	7.9	25
Tobacco	5.1	78
Cotton	7.8	61
Peaches	1.9	39
Oranges	2.0	38
Dried raisins	2.6	56
Milk	8.9	63
Beef meat	2.8	61
Pork meat	3.3	33
Eggs	2.5	59
Poultry	2.9	57
Sheep and goat meat	6.2	35
Total crop production	71.6	36
Total livestock production	28.4	48
Total agricultural production	100.0	40

Source: Bazioti, G. and Bourdaras, D. 1990

Agricultural holdings in the Mediterranean States of the South, namely Portugal, Italy, Spain and Greece receive much lower amounts of compensation from the Guarantee section of EAGGF in absolute terms varying from 4300 ECUs for Greece to 1000 ECUs for Portugal (table 5.3).

It should be pointed out that EAGGF's Guarantee expenditures are distributed quite unevenly as a small percentage of large agricultural holdings account for a high share of total support.

In general, a substantial increase in trade between Greece and the rest of the Community was the effect of liberalization practices that prevailed between the two regions. Greece, by adopting the external trade regime of the European Community, essentially raised very effective trade barriers with many but not all of the countries in the rest of the world.

The country's trade balance deteriorated substantially, from a net trade surplus with the Community's member states and a net trade deficit with third countries to a large and growing net trade deficit with the Community and a net trade surplus with third countries. This pronounced change in the trade pattern is a clear indication of trade diversion (Sarris, 1984). The adverse developments in the balance of agricultural trade are due to the increase of imports of meat and dairy products and the dramatic change of the pattern of meat imports. All predictions for the growth of total net exports of fruit and vegetables did not materialize.

Table 5.2

EAGGF (Guarantee) section expenditure as a percentage of total value of agricultural production by member state (1987, 1990, 1993, 1994)

Member State	1987	1990	1993	1994
Belgium	15.2	14.4	19.8	17.1
Denmark	16.9	16.3	20.6	20.0
Germany	15.5	15.7	15.7	16.5
Greece	19.6	24.0	33.1	31.2
Spain	2.8	8.4	19.1	19.9
France	13.4	10.6	19.4	18.2
Ireland	26.1	39.8	38.3	34.4
Italy	11.2	11.4	14.4	10.7
Luxembourg	0.9	2.7	3.9	6.5
The Netherlands	19.5	18.3	14.7	11.4
Portugal	---	6.1	15.3	22.0
United Kingdom	10.4	10.5	15.7	16.5
EU12	12.9	13.1	18.2	17.2

Source: *The agricultural situation in the Community,* European Commission (various issues)

Such trade developments actually imply heavy invisible transfers from Greek consumers to the Community's producers. Such a net transfer must be compared with the clearly visible and substantial net transfers from the Community's budget to Greece. The country's accession did have direct consequences for Greek agricultural producers as well those whose income position was directly related to the prices of the products they produced as well as the prices they paid for all the necessary purchased means of production.

During the decade before accession real producer prices increased for practically all crop products except for industrial crops while real prices of livestock declined. After accession, the relative price structure within Greek agriculture was profoundly changed. Price increases were higher for industrial crops and animal products (Sarris, 1984). Throughout the post accession period and until recently, Greek agriculture did not manage to capitalize on its comparative advantage by promoting the production of Mediterranean crop products. It seems that the price and aid system that prevailed was not sufficient to assist domestic agriculture in overcoming structural weaknesses generally manifested by the small size and fragmentation of holdings, the inefficient marketing system and infrastructure. In essence, the income position of Greek farmers does not seem to have improved in spite of the substantial income support from both Community and national sources. The weak border control for Mediterranean products exposed Greek farmers to international competition immediately after accession that necessitated increasing funds for income support. Efforts to modernize agricultural production structures in Greece proved insufficient.

Table 5.3
EAGGF (Guarantee) section expenditure per holding by member state
(1987, 1990, 1993)

			'000 ECUs
Member state	1987	1990	1993
Belgium	10.4	11.0	17.4
Denmark	12.3	13.0	18.1
Germany	6.0	6.5	8.1
Greece	1.9	2.8	4.3
Spain	0.4	1.4	3.0
France	6.2	5.6	10.3
Ireland	4.4	7.7	9.8
Italy	2.0	2.1	2.7
Luxembourg	0.4	1.3	2.1
The Netherlands	23.3	24.5	19.8
Portugal	0.4	0.6	1.0
United Kingdom	7.2	8.1	11.6
EU12	3.3	3.8	5.5

Source: *The agricultural situation in the Community,* European Commission (various issues)

Structural improvements are partially financed by the Guidance section of EAGGF along with other Community and national funds. Structural policy instruments have provided assistance for improving farm structures in Greece with an aim to create a more competitive and viable agricultural sector.

However, funds allocated by the Guarantee section of EAGGF for the implementation of price and market support measures have been far more substantial. The structural part of the policy has become less effective than market price policy which has been more decisive in determining the final outcome as far as competitiveness is concerned.

The EAGGF's Guidance section expenditures in Greece are above the Community's average. In 1993 they exceeded 600 ECUs per holding or 100 ECUs per hectare of Utilized Agricultural Area (UAA) as compared to an average of about 500 ECUs and 26 ECUs for the Community respectively. Guidance expenditure per hectare in Greece is the highest in the Community (table 5.4).

Table 5.4
EAGGF Guidance section expenditure per holding and unit of utilized agricultural area (UAA) by member state in ECUs
(1987, 1993)

	Per holding		Per unit of UAA (ha)	
Member state	1987	1993	1987	1993
Belgium	267.5	558.7	14.9	31.0
Denmark	134.9	271.7	4.1	7.3
Germany	181.9	564.8	10.2	20.5
Greece	149.9	637.7	18.3	113.2
Spain	51.5	300.8	2.9	16.9
France	267.2	795.1	7.7	22.5
Ireland	444.9	986.9	17.0	37.3
Italy	48.6	360.1	5.5	43.1
Luxembourg	972.2	2,571.4	30.6	70.3
The Netherlands	117.9	166.2	6.8	9.7
Portugal	161.9	643.6	13.7	86.55
United Kingdom	344.6	421.3	4.5	6.0
EU12	135.5	491.2	7.3	26.0

Source: *The agricultural situation in the Community,* European Commission (various issues)

Price policy, implemented in Greece, has had an impact on the competitive position of the agricultural sector. Prices supported by the CAP, just as in any other member state, provided incentives and disincentives that resulted in production adjustments and kept marginal producers in business in the short run. Figure 5.1 demonstrates that the implementation of the CAP in Greece for the five year period between 1989 and 1993 has been more effective than in any other member state, except the United Kingdom, in prolonging the operation of marginal agricultural holdings as the total number of holdings was reduced by only 3% approximately.

By safeguarding the engagement of small and marginal agricultural holdings in the production process, the CAP has had a positive influence in Greece as far as employment is concerned. Adjustments toward improving farm structures have been very slow, contrary to developments in Portugal and Spain, for example, where the number of agricultural holdings was reduced in the same period by 18% and 13% respectively. The CAP therefore has had a positive influence on agricultural employment in Greece but has not been as effective in the development of a more competitive farm structure.

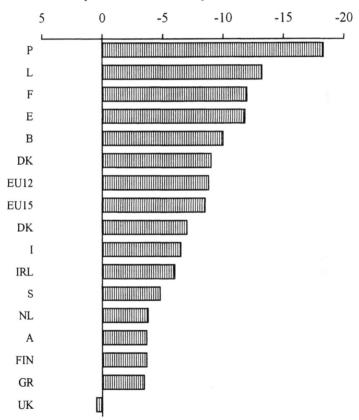

Figure 5.1: **Changes in the number of holdings between 1989-90 and 1993 (%)**

Source: Eurostat

151

It is generally acknowledged that the CAP benefits are greater for products such as dairy and cereals than horticultural products and fruit. Greece, as a Mediterranean member state would tend to benefit less than the northern and richer member states as a significant part of the national production enjoys a lower degree of support.

Stronger support for livestock and other products did not result in a desirable development of the sector due to a number of natural, structural and institutional constraints (Zanias, Maraveyias, 1996). The fact that the number of farms in Greece has been reduced at a relatively low rate could be partially explained by the higher level of support attached to a number of products of national interest such as cotton, olive oil, tobacco and durum wheat.

Table 5.5 sums up the total impact from the implementation of the CAP in Greece for the year 1993. It is demonstrated that the overall effect is a positive one as net financial transfers to Greece corresponded to 2.8% of the country's Gross National Product (GNP). The net effect on income is 1.6% of the GNP, taking into account the loss of surplus for the consumers and the financial burden for the Greek taxpayer.

Table 5.5
Transfers, benefit and cost from the implementation of the CAP in Greece (1993)

	Financial transfers	Amount (in million ECUs)
1.	Transfers from the EU	2,817
2.	Transfers to the EU	485
3.	Net financial impact (1-2)	2,332
4.	Impact from trade	-179
5.	Net transfer (3-4)	2,153
	Net Transfer as a percentage of GNP	2.8
	Impact on income	
6.	Benefit to producers	3,346
7.	Benefit to consumers	-1,635
8.	Cost to taxpayers	485
9.	Net social benefit (6+7-8)	1,226
	Total impact on income as a percentage of GNP	1.6

Source: *Agriculture in Europe. Arguments in favor of a radical reform*, Ministry of Agriculture, Fisheries and Food of the United Kingdom 1996

The evolution of farm income and its distribution

During the period between 1981 and 1994 agricultural income per annual working unit (AWU) in Greece was moderately increased in real terms by an average annual rate of 1.7%. In 1995 real income per AWU was increased by 2% to fall again by approximately 1.2% in 1996 (provisional estimate). The rate of increase of agricultural income in Greece, for the same period exceeded that of Belgium, Germany, France, Ireland, Italy, Portugal and the United Kingdom (Eurostat, various issues). It should be noted however that in absolute terms, farmers' incomes in Greece are the lowest in the Community.

The composition of the total incomes of Greek households whose main income stems from farming is presented in figure 1.7 as an average over two-three year periods, that of 1982-84 which coincides with the early accession years and 1988-90 which refers to a time ten years later. During the first period, Greek agricultural households received just over half of their total income from farming. In the later period income from farming constituted less than 40% of total income. The second most important source of income of agricultural households was "other independent activity" i.e. self employed activity such as operating other non-agricultural businesses for the period 1988-90 and wages for the period 1982-84. Other sources of income by order of significance were property, 4% and 12% for the years 1982-84 and 1988-90 respectively, social benefits (8% and 7%) and imputed rental value (3% and 3%).

Among all the EU member states, Greek agricultural households in aggregate, received the highest proportion of their income from sources other than farming. Income from other independent activities is also the highest when compared to the situation in any other member state. Income from property had expanded substantially during the 1980s. It should be pointed out that income from property increased four times over the period under investigation.

Due to the low agricultural income in absolute terms, national policy does not impose any income tax on agricultural households. In Greece, less than 1% of total income is taken by taxation from agricultural households, corresponding to the lowest taxation rate in the entire Community.

Agricultural households in Greece seem to compare favorably with all households in terms of the average disposable income they possess just as in all member states with the exception of Portugal. The total disposable income for agricultural households is slightly higher than the all household average (Hill, 1996). Contrary to prevailing convictions and in spite of the general low income problem in the country, agricultural households, in aggregate, do not comprise a particularly disadvantaged group in terms of their average disposable income. They are subject to large short term income variations

(Hill, 1996). Total income of agricultural households in Greece and, in particular, agricultural income in the country is unevenly distributed. In relative terms, Greece has the highest number of farms in less favored areas. Approximately 60% of all farms in Greece are located in such areas, a third of which are located in mountainous areas (Taminga, 1991).

Farms in the less favored areas have lower income than farms in the rest of the country due to low productivity and the less developed regional economy, in spite of higher compensation per unit of cultivated area or animal unit. The income gap between normal areas and less favored areas is not offset by the agricultural policy measures that have been applied before and after the country's accession into the European Community. Low output prices have prevailed in combination with relatively higher prices of inputs. The relatively high proportion of farms engaged in conventional, low income farming types is an additional factor that explains income behavior in less favored areas. Production per unit of cultivated area and consequently gross margins play an important role in explaining income differences.

It seems that regional development, in particular, plays an important role in explaining agricultural backwardness in Greece; regional policies pursued have not proved effective in solving farmers' income problems. Moreover, market structure policy actually implemented in Greece has failed to tackle the low output-high input price problem, especially in less favored areas. General structural and macroeconomic policy measures have proved ineffective in bringing about productivity gains, making compensatory allowances indispensable.

Market and price policy have been, by far, the most important policy for income support. Socio-structural policy has played a role of secondary importance. Table 5.6 demonstrates that economic performance (GDP per capita), as well as added value in agriculture per agricultural labor unit are not necessarily directly correlated with the CAP transfers as determined by the Common Organization of the Markets. For example, in Attica and Central Greece where both GDP per capita, as expressed in purchasing power standards and value added in agriculture per agricultural labor unit, are among the highest CAP transfers and are inferior to those observed in a number of other regions such as Eastern, Central and Western Macedonia as well as Thessaly. Similarly, Crete is characterized by low CAP transfers but high added value in agriculture. Many farms in Crete have high production output, therefore farms are benefiting more from the market and price policy than farms with limited production elsewhere in spite of high CAP transfers.

It has been stated previously in this text that the majority of Greek farmers are operating small agricultural holdings. They are a diverse group ranging from part time farmers who maintain off farm activities to full time operators. For small farmers who earn insufficient incomes from farming, traditional farm policies implemented in Greece during the last decades have not been successful in offering the assistance required. Off farm income however has recently been enough to give the average farmer a total income that is higher than the income of the average household.

Various policies have distributed income quite unevenly. Results of large scale research conducted in 1987/1988 (Damianos et al., 1994) demonstrate a high degree of heterogeneity among agricultural households throughout the country. The unequal distribution of income suggests that farms can be divided into distinct subgroups on the basis of two factors, namely, age and the extent of farm employment.

These two factors provide insight into operators' reliance on farming for their income, the extent to which they are likely to suffer low incomes, the alternatives available to them for improving their on and off the farm income, the appropriateness of policies implemented and the kind of policies by means of which income problems could be addressed more effectively.

Table 5.6

Indicators of economic performance and agricultural transfers by region

Region	GDP per capita PPS,[a] 1991	CAP transfers per ALU,[b] 1994	Added value in agriculture per ALU
Eastern Macedonia	43.6	128	88
Central Macedonia	45.4	141	110
Western Macedonia	46.6	134	164
Thessaly	43.4	175	134
Epirus	36.5	65	43
Ionian Islands	41.3	49	61
Western Greece	40.9	76	90
Central Greece	57.5	110	95
Peloponnese	47.2	75	91
Attica	53.8	113	285
Northern Aegean Islands	34.9	50	59
Southern Aegean Islands	47.0	62	76
Crete	43.0	80	136
Greece		100	100
EU12	100.0		

[a] PPS: Purchasing Power Standards
[b] ALU: Agricultural Labor Unit

Sources: Tarditi, S., *Impact of the agricultural price policy on EU cohesion. A regional analysis,* European Commission, 1996

155

Three major categories have been defined. Category A refers to pluriactive agricultural households whose managers are less that 65 years old and devote more than 50% of their working time to activities off the farm. Category B refers to "full time" agricultural households whose managers are less than 65 years old and devote more than 50% of their total time to activities on the farms. Farms whose managers are over the age of 65, regardless of whether they work off the farm or not comprise category C.

Approximately 20% of all farms in Greece were managed by farmers older than 65 years in 1987/1988. Such farms tend to have the lowest total household income much of it coming from social security and other payments. Farm households of this category tend to rely on public income support policy measures. These senior operators do not seem to have very many things in common with young commercial farmers. However, agricultural policy measures have been attempting unsuccessfully to raise incomes of their families by applying identical policy instruments.

About 25% of the total number of farms in Greece have "part time" operators in 1987/1988 (category A). Such pluriactive farmers, on average, have lower incomes as compared to full time farmers but higher incomes than farmers over the age of 65, most of it derived from off farm employment. In 1987/1989 off farm jobs provided pluriactive farm households in Greece with over 70% of their total income. Thus, conventional farm policy measures can have an impact on 30% of such households' total income, leaving the rest, 70%, unaffected. "Part time" farmers tend to operate smaller farms. For this category of farmers, farming plays a supplemental role in the family's well being. Where "part time" farmers do have income problems, policy measures other than conventional agricultural policy instruments in the non-farm sectors are likely to be more appropriate.

The remaining 55% of farmers were defined as "full time" operators. Off farm incomes do exist but are typically lowest on these farms fluctuating around 15% of total income on the average. Farming activity is more extensive and is more likely to generate positive returns as it is favored by conventional agricultural policy measures. The total income of "full time" operator farm households average the highest of the three groups. This is the group of farmers who are best served in relative terms by policies incorporated in the Common Organization of the Markets. Yet, even within this group, income differences are large enough to call into question the effectiveness and appropriateness of typical farm policy measures as provided by the common price and income policy and the common sociostructural policy.

In essence, income problems characterize a good part of farm households in all the three categories. Among farmers of categories A and B, age is an important influence on a farmers' range of options for improving either farm and non farm income as they can take better advantage of any alternatives available. The longer work life expectancy characterizing farmers of categories A and B make education and training to increase on and off farm income more attractive than for older managers. In the short run, many younger farmers in both groups face difficulties in improving either farm or non farm income. In the present depressed agricultural conditions, farm income is especially difficult to improve. In generating off farm income, farmers face serious difficulties when operating in local economies that are highly dependent on agriculture.

Older farmers, whether in group A or B are faced with difficulties that can become insurmountable. These farmers are often not in a position to take full advantage of retraining alternatives, more promising farm practices or non farm jobs. In general, farmers who continue to operate a small farm, in particular when they are in their middle years are unlikely to expand or diversify operations, especially in the current restrictive agricultural and economic environment.

Figure 5.2 displays the distribution of farm household income in Greece. It becomes apparent that total income for the various categories stemming from all possible sources as a result of the initial distribution of assets among households and policies pursued, are unequally distributed as indicated by the Lorenz curve and Ginni coefficients. Five percent of all agricultural households in Greece account for about 27% of total income for all households whereas 10% of all households account for about 37% of total income. Half of total income is accounted for by only 20% of agricultural households. Total income inequality is higher for "full time" agricultural households. The high degree of concentration is partly due to agricultural price and income policies pursued. Only 4.4% of total income is accounted for by the "poorest" 20% of all households.

Figure 5.3 presents the distribution of the gross agricultural income for all households and households in categories A and B. Just as in the case of total agricultural households' incomes, gross agricultural income is distributed more unequally for "full time" households. It is therefore concluded that in spite of the high expenditures attached to agricultural policy implemented in Greece during the last few decades, low income problems in absolute terms continue to characterize the majority of agricultural households in the country. The larger agricultural holdings account for the lion's share of expenditures for the implementation of agricultural policy programs.

In sum, rural households in Greece are particularly diverse with respect to their economic activity, economic welfare and production whereas they depend on farm activity for their incomes in various degrees. A better understanding of forces that play a decisive role in income generation is therefore indispensable.

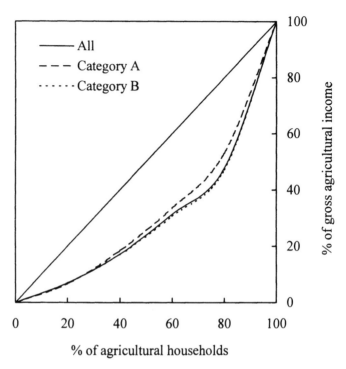

Figure 5.2: **Lorenz curves indicating agricultural households' gross agricultural income distribution**
Source: Damianos et al., 1994

Income generation is more unequal for rural households that depend heavily on farming as compared to the rest of the households in rural Greece. Each household's given volume of agricultural production or net income from farming is a decisive determinant of income inequality especially for category B farmers. Income inequality estimates are underestimated as they do not take into account income stemming from the capitalization of agricultural policy measures on the value of land. Because of the absence of zoning restrictions as well as property and income tax for practically all rural households, agricultural land values in Greece have been maintained at the highest level in the entire Community. Policy measures aiming at leveling off income inequalities among agricultural households, the design of agricultural land policy along with the development of employment opportunities off the farm could be instrumental in securing the fragile viability of rural communities throughout the Greek countryside.

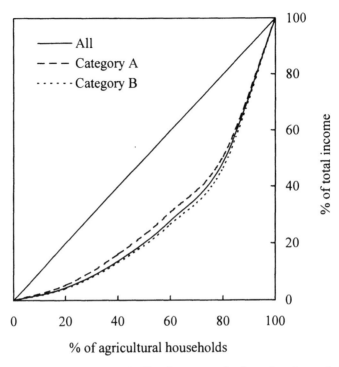

Figure 5.3: **Lorenz curves indicating agricultural households' total income distribution**
Source: Damianos et al., 1994

Characteristics of employment

Employment in Greek agriculture has been decreasing at relatively low but increasing rates. Agricultural policies pursued during the last decades as well as policies for rural development and general macroeconomic policies have been decisive in shaping demand and supply conditions for labor in agriculture and the non agricultural sectors throughout the country. The annual rate of change in agricultural employment in Greece was -2.3% during the 1970s. During the first decade after accession this rate was reduced to -1.3%, the lowest in the Community. In the case of the Netherlands where agricultural employment was increased at an annual rate of 1.7% and the United Kingdom where employment in agriculture decreased by 1.2%.

During the recent five year period between 1990 and 1994 employment in Greek agriculture decreased at an average annual rate of 2.9%, that is substantially higher than rates experienced in the preceding 20 year period. The higher rate of decrease in agricultural employment after 1990 constitutes an indication that senior farmers are abandoning agricultural activities at increasing rates, leaving no successors to take over farming. In general, Greece has a lower average annual rate of decrease in agricultural employment than countries in the Union that are favored by a much higher degree of financial support per holding, as expressed by EAGGF's expenditures.

Taking into account that about 30% of the heads of agricultural households in Greece are over 65 years of age, it is expected that the reduction of employment in farming will move at a much faster rate in the immediate future. Approximately two kinds of the total number of operators of agricultural holdings in Greece are underemployed in agriculture as they work on the farm less than six months during the year while the percentage of those occupied on a full time basis does not exceed 15% as compared to a Community average of 30%. Out of the two thirds of operators who work less than 6 months a year on the farm, about 60% do not have any other occupation. Out of all operators who work more than 6 months a year on the farm almost 90% do not supplement their incomes by off farm activities.

It has been stated earlier in this chapter that agricultural policy measures applied in Greece since the mid 1970s have been effective in maintaining agricultural incomes per labor unit at approximately the same level or more precisely, farm policy has led to a slight average annual increase of incomes. However, the fall in agricultural prices observed during recent years is expected to bring about a fall in farm incomes which in turn will encourage farmers to seek employment opportunities off the farm.

The agricultural community in the country is expected to be deprived of its most energetic and dynamic members. If no countervailing measures are taken, the age structure will deteriorate even further and so will the competitive position of the country's agricultural sector. Human resource characteristics in rural Greece, namely age and level of education hinder agricultural development as well as rural development in general especially in the less favored areas, the mountainous and coastal areas and the islands of the country. Human resources available in such areas, currently employed in agriculture are not capable of supporting any effort for economic development except for satisfying the demand for unskilled labor in other sectors.

Agricultural policy and policies for rural development, although not very effective in Greece, have been aiming at facilitating the process of structural adjustment and the reduction of employment in agriculture. For Greece, the time has come when policies should be adjusted to encourage young, better educated and trained farmers to enter the agricultural sector.

Factors influencing policy success

The performance of agricultural policy and the achievement of objectives

The overall impact of specific national and community agricultural policies and policies for the rural sector implemented in Greece during the last decades cannot be easily identified. What would the outcome be without such policies is more or less presumptive, as no general consent can be reached concerning an appropriate reference system that would permit explicit and incontestable "with" and "without policy" comparisons (Koester, 1984).

It is difficult to attribute any modest labor and total factor productivity gains, which have had reasonable objectives achieved, which are due to market and price support, the main instrument of policies implemented. Furthermore, any improvement in the standard of living of the farming and rural populations can not be ascribed to price support. In addition, income stemming from agricultural activities, on the average, contributes to total rural households' income by less than fifty percent and is unequally distributed among agricultural holdings. It is doubtful whether price support has actually increased agricultural and hence, total incomes of farms in rural Greece in the long run.

Price increases have certainly not improved income conditions of small farmers, the great majority of farmers in the country. Larger and wealthier farmers have been favored however by policies centered around price support, the core policy instrument. Market stability and other stated objectives of national but mainly Community agricultural policy has not been attained at the national level. In particular, the degree of self sufficiency in livestock production and other products for which Greece has not been self sufficient has not increased, whereas the negative gap in the balance of trade of agricultural products has been maintained at high levels for a long period of time and is seriously affecting the country's balance of payments. On the other hand, producers' price increases have resulted in high consumer prices. In general, Greece has encountered serious difficulties in pursuing national policy objectives due to the inadequate division of jurisdiction between the European Community and the state. Income distribution for example has been very much affected by the Common Market and Price Policy, which has eroded structural policies pursued at the national level in the context of the common framework of sociostructural measures.

The European Community has not claimed responsibility for total Community welfare as the Treaty of Rome avoided any commitment for the achievement of economic and social cohesion among member states. In

Greece, structural adjustment and change has been slow. For various reasons, farmers have been able to stay in agriculture by accepting lower farm incomes at a time when structural change in northern member states has already taken place to a great extent.

Well structured, competitive farms constitute a small part of the total number of farms in Greece. Accelerated structural change is not easy to come about due to acute difficulties and constraints such as low employment opportunities in the non agricultural sector, competition for land, especially in better quality irrigated areas, the immobility of land resources due to its retention in the expectation of capital gains (Bergman, 1984).

Especially during the initial stages of the harmonization period with the European Community, structural measures were not integrated into a comprehensive rural development package addressed to all economic sectors and adjusted to the specific characteristics and needs of particular regions. Structural improvements were attempted at the local level by some distant bureaucratic structures in Brussels and Athens. Later on, when programs were inspired and designed at the regional level they lacked rationality, coordination, and compatibility with goals initially set. Integrated programs were implemented that did not have the effect expected in increasing farm size or in restraining the inflated land values. A major handicap has been the low effectiveness of institutions entrusted with the support of the complex agricultural activity. The weak performance of institutions has jeopardized the effectiveness of Community transfers in support of common policies.

The bias in the CAP support in favor of larger farms continued until after the MacSharry reform as the bulk of net benefits still goes to the large cereal, milk and beef farmers. Consumer gains were substantial and compensated for any net loss in producers' surplus accrued to small and medium size farms.

It is worth noting that Greece along with other southern member states has assumed high administrative costs for implementing agricultural policy measures especially after the 1992 reform, due to the large number of farms involved (Sarris, 1994).

Long term policy issues and challenges

Developments in the external economic and trade environment have already had direct consequences for the CAP as portrayed by the radical 1992 reform. The international dialogue on free trade, to be continued under the auspices of the World Trade Organization will have implications that are likely to extend in the long run. European agriculture will increasingly have to depend for its share of the market, on its ability to compete effectively as agricultural policy arrangements will eventually cease to provide the protective shield to farming currently in effect (Marsh, 1993).

Prospective international developments, namely growing regional integration, globalization of food markets, rapidly growing trade, especially in processed products, advancements in biotechnology, transportation and communications, growing market power in the food processing and distribution industry have intensified the pressure for policy reform. The trade policy environment is a principle factor that will affect the face and the direction of adjustments faced by farmers at the Community and at the national level. Agricultural policy will be better targeted, more cost efficient and decoupled from farm production.

The debate for a new European agricultural policy is now starting in the framework of the comprehensive Commission's proposals entitled "Agenda-2000". According to the Commission's plan, price policy under the CAP is to be further linked to international price trends while at the same time structural policy is being intensified. The European cereal, milk and beef producers would receive internationally based prices for their products. The proposed change might have a negative short term effect on the incomes of many European farmers. However, income supplements differentiated from region to region will continue to be in effect, in addition to what farmers receive in the form of market prices. According to the proposals, specific income supplements could be granted at the regional level for the production of public goods such as natural assets, quality of the landscape and the environment. Farmers throughout Europe will be recognized as preservers of nature and the landscape. The proposed measures could become a major source of income for Greek farmers given the very many areas of special ecological value and the extent of less favored areas in the country.

The incorporation of a strong regional component in agricultural policy as well as a genuinely integrated approach to rural problems will create opportunities for domestic agricultural and rural development. Such a policy designed and pursued at the level where all actors are directly related to problems concerned will be a more effective policy with respect to regional problems such as the enhancement of structures and employment. Solutions to income, employment and development problems will therefore be traced and detected on the basis of specific regional features and diversity (Woltjer, 1993).

Although the new policy will not reverse the general trend toward less direct employment in agriculture, employment possibilities could be promoted by national policy on business creation in rural areas. Rural areas that will continue to suffer major job losses in farming will have to exploit their local potential for development. The targeted use of funds could thus be used efficiently in order to create new and worthwhile employment. In order for rural areas in Greece to prosper in new areas and activities that are linked to real demand, markets will have to be developed.

163

A comprehensive and coherent set of interventions, in the framework of a Common Rural Policy is required to reverse the decline of areas which are at a distance from large population centers and base their economies exclusively on small scale farming activities.

National development plans and operational programs in the context of the Third Community Support Framework to be launched right after the expiration of the century, should be developed and implemented taking into account the role of all complementary sections of the economy. The combined set of policy measures for the development of rural areas should be coherent and relevant to clearly defined and realistic goals for each of the country's diverse regions. The capacity and competence of the appropriate institutions, operating at the national level should undoubtedly be enhanced in order for the country to benefit from Community support envisaged for rural development. (Mannion, 1996a)

The importance of support provided to Greek agriculture and the country's rural areas from the EAGGF's Guarantee section should not be underestimated. Its contribution for the implementation of conventional sectoral measures is still vital, particularly in areas specializing in certain production types such as cotton, tobacco, olive oil, etc. as well as in areas where non agricultural activities are limited. Price and income support provided by EAGGF should allow for the necessary flexibility required in order to take into account the vulnerable character of such areas, where the establishment of non farming activities is difficult or impossible to come about in the short run (Mannion, 1996b).

This territorial rather than sectoral orientation for the implementation of rural development programs concentrated where they are most needed, incorporated within a distinct rural policy framework constitutes a Community policy that best fits the situation in Greek agriculture and countryside. Making Greek agriculture more efficient and competitive requires facilitating the transfer of resources to new uses. Sound macroeconomic policies that are already in force are bound to create rewarding employment that could add to the country's wealth.

Some of the critical areas around which rural policy objectives should be developed are training policies that would enhance human resource capabilities, both in the farm and the non-farm sectors, the development of rural transport and infrastructure so that access to employment for rural citizens is increased, the encouragement of rural enterprises by assisting local entrepreneurs, the provision of rural advisory services by expanding and improving the extension network, the provision of the required information by all available means including electronic communication (Marsh, 1993).

The general improvement of the quality of life by promoting positive environmental and countryside stewardship measures that improve the sustainability of rural areas making them more accessible and attractive to urban citizens should be a priority goal. Structural adjustment, social stability, cohesion and economic growth could then be facilitated.

Bibliography

Bazioti, G. and Bourdaras, D. (1990), *The Support of Greek Agriculture in 1988*, Unpublished Report, Ministry of Agriculture: Athens. (in Greek)

Bergmann, D. (1984), 'North-South Balances and Conflicts within the Community', Paper presented at the *Workshop on the Reform of the CAP*, University of Sienna, Italy, February 17-18.

Damianos, D., Demoussis, M., Kasimis, C. and Moysides, A. (1994), *Pluriactivity in the Agricultural Sector and Policies for Rural Development in Greece*, Foundation of Mediterranean Studies: Athens. (in Greek)

European Parliament (1995), *Incidences Financieres du Fonds Europeen d' Orientation et de Garantie Agricole en Grece, de l'Adhesion (1981) a la Creation du Marche Unique (1992)*, Serie: Agriculture-Peche-Forets, W-26/I, Direction Generale les Etudes: Luxembourg.

Hill, B. (1996), *Total Income of Agricultural Households, 1995 Report*, Eurostat: Statistical Document, Luxembourg.

Koester, U. (1984), 'The Role of the CAP in the Process of European Integration', Paper presented at the *Workshop on the Reform of the CAP*, University of Sienna, Italy, February 17-18.

Mannion, J. (1996a), 'Situation and Trends of the National Policies for Rural Development', Paper presented at the *European Commission Conference on Rural Development*, Cork, Ireland, November 7-9.

Mannion, J. (1996b), 'Strategies for Local Development in Rural Areas. The Bottom Up Approach', Paper presented at the European Commission Conference on Rural Development, Cork, Ireland, November 7-9.

Marsh, J. (1993), *The Future of Agriculture in Europe*, Unpublished Comment, University of Reading.

Sarris, A, (1984), 'Agricultural Problems in EC Enlargement. The Case of Greece', Paper presented at the *Workshop on The Reform of the CAP*, Sienna, Italy, February 17-18.

Sarris, A. H. (1994), *Consequences of the proposed Common Agricultural Policy reform for the Southern part of the European Community*, European Economy Reports and Studies, No. 5.

Taminga, G. F., Von Meyer, H., Strijker, D., Godeshalk, F.E. (1991), *Agriculture in the Less Favored Areas of the EC-10*, Agricultural Economic Research Institute (LEI-DLO), The Hague.

Woltjer, E. (1993), 'Steps Toward the Renewal of European Agricultural Policy', Paper presented at the Conference of the Group of the Party of European Socialists *The Future of Agriculture in Europe*, Strasbourg, December.

Zanias, G., Maraveyias, N. (1996), 'The Impact of the CAP on Cohesion Among EU Member States', Paper presented at the International Conference on *Economic Integration in Transition*, Athens University of Economics and Business and York University, Athens, August 21-24.

6 International developments and perspectives for Greek agriculture

The CAP reform and the Uruguay Round

The CAP and the international environment

During the last fifty years world trade has been rapidly growing, resulting in market globalization and in increasing interdependence between national economies. Thus, unilateral decision making and action taking may have serious implications for international markets. International economic integration implies a wide exposure of national economies to foreign competition as well as a need for continuous structural adjustments by agricultural societies and national economies along with compliance of national policies with international provisions. The special characteristics of agriculture (activity susceptible to a variety of external factors, inelastic supply of and demand for agricultural products, low income elasticity, need for highly specialized factors of production, etc.) raise the uncertainty related to the production process. Consequently, agriculture is a sector that usually receives attention by policy makers and becomes subject to government intervention, aimed at guaranteeing a minimum income level for farmers, safeguarding adequate food supplies, and stabilizing the markets. Such were the actual Common Agricultural Policy (CAP) objectives at the time of its establishment. Other goals were also put forward such as an increase in productivity by promoting the rational use of resources and the guarantee of reasonable prices to consumers. The main mechanism used under the CAP framework was the common organization of the markets and the market price support system.

Meanwhile, international organizations set a new framework for policy making in agriculture. During the Uruguay Round, the General Agreement on Tariffs and Trade (GATT) concluded that policies should be adjusted toward greater market orientation, transparency, and discipline. Objectives such as rural development and environmental protection received attention, especially in relation to the role that agriculture plays. In addition, attention was paid to the role of less developed countries, in satisfying their nutritional needs and speeding up structural adjustment. In general, the need for continuous structural adjustments in the agricultural sector was widely recognized. Although the orientation of the common organization of the markets varies according to product groups, the common element of the support system was the specification of guaranteed prices. Most of the products such as cereals, sugar, dairy products, meat, were subject to market intervention complemented by the external measures of export subsidies and variable import levies. The operation of the CAP resulted in problems relating to the CAP's failure to meet its objectives and in problems created by the operation of the CAP itself. With regard to the first category of problems, the CAP:

- was proven unsuccessful in exploiting the benefits of comparative advantage, since high support prices and lack of common prices[1] led to the misallocation of resources,[2] thus prohibiting the formation of specialization patterns on efficiency grounds in the European Communities;
- failed to help the income of poor farmers, since subsidization per unit of output benefited larger enterprises which could expand output more easily;
- fell short of reducing urban and rural disparities in individual earnings;
- failed to support the structural reform programs in promoting rural development and in encouraging the modernization of farm structures and of production modes;
- effected structural imbalances between the supply and demand for agricultural products; and
- failed to adjust the quality and the composition of agricultural production to changes in consumer preferences.

With regard to the second category of problems, the CAP:

- created structural imbalances and enormous budgetary expenditures;
- had a depressing effect on the world prices of agricultural products, especially of grains and dairy products causing an increase in both the social cost and the cost of financing export subsidies;
- had a destabilizing effect on world markets by making the EC trade functions more inelastic. As the CAP prevented any quantity adjustments

to shocks by the EC, it transmitted any disturbance to the world market, which would have to absorb changes through higher price adjustments;
- widened the inequalities in the distribution of income within the agricultural sector and between the agricultural and other sectors of the economy; and
- created destabilizing conditions for the economies in transition or in development stages.

It became clear that the market price mechanism was impossible to lead to a certain level of farm income without creating structural market imbalances, increasing budgetary pressures, problems to the world markets, and friction with large trade partners, such as the United States.

The CAP reform

In 1992 a political agreement was reached among the twelve member states of the European Union on changes to the Common Agricultural Policy reflecting a more radical reform of market mechanisms which had been undergoing changes since 1984. Although the reform was based on the 1962 principles of market unity, community preference and financial solidarity, it did add new objectives. The system of support for agricultural income, mainly through guaranteed prices was replaced by a system of direct income compensation. The reformed Common Agricultural Policy also incorporated a number of other innovations. Income compensation was matched with a policy of production controls. Set aside for arable crops became mandatory for large farms as a means to control production increases. Income compensation was calculated on grounds of a historically determined basic area in order to discourage further increases in production.

The reform of the Common Agricultural Policy was put into practice as of the marketing year 1993/1994 and did not involve all the sectors subject to a common organization of the markets. It was limited initially to the main crops and animal products plus tobacco. Future proposals were planned for other major sectors such as wine, fruit and vegetables and sugar. In the case of arable crops the reform was based on a gradual but substantial reduction in the guide price for cereals so as to bring cereal prices closer to the world price level. Any resulting loss of income was offset by compensating aids provided, calculated on a flat rate basis per unit of cultivated area depending on average regional yields.

The common organization of the markets for tobacco, an important product for Greece, was also incorporated into the 1992 reform. The various variety groups were reduced to a minimum. Maximum guaranteed quantities for each particular group were also reduced. In the dairy and beef sector, reductions of milk quotas and prices were imposed. Any income losses caused by these changes were to be offset by the payments of premiums per head of cattle provided extensification practices were implemented. Extensification was achieved by limiting the number of animals in relation to the area producing fodder. In the sheep sector, premiums were attached to grazing land which diminished according to the size of each flock. In general, smaller farms are relatively favorably treated with respect to aid granted. The Council also enacted a number of accompanying measures of structural character to the 1992 reform of the Common Agricultural Policy in order to make it more agreeable to producers throughout Europe. These measures aim at encouraging the early retirement of farmers, promoting afforestation of farm land and supporting the introduction and adoption of environmentally friendly methods of production.

The reform will undoubtedly affect farmers, consumers and taxpayers in Greece just as in the rest of the Union. Farmers were expected to be affected by the substantial fall in market prices. Such a fall was expected to have a negative impact on producer prices and agricultural incomes which was to be offset by direct aids granted. These in turn were to vary sharply from farm to farm depending on commodities produced, size and location. However, over the last few years since the 1992 reform agricultural incomes in the Community developed favorably (European Commission, 1995). Competitiveness of cereals and other European products was improved in the Community as well as in export markets. As prices of the primary production upon which final food products are based fall, consumers are in a better position as long as price reductions affecting producers are passed on fully to the former.

Since the 1992 CAP reform, a number of additional major agricultural policy changes of importance for Greece have taken place. Motivated by the need to find additional budgetary resources to cope with the beef crisis (BSE-Bovine Spongiform Encephalopathy) the Council agreed on a permanent reduction of payments under the arable reform scheme. In addition, a reform similar to the cereals sector has been agreed by introducing a reduction in the intervention price for rice for 1996/1997 with compensations and penalties for national areas that exceed limits within areas which they can produce without incorporation of set aside measures.

In the fruit and vegetables sector, reform of policy has been agreed from 1996 onwards. The support mechanism of product withdrawal from the market has been substantially reduced and financial assistance to producer organizations with incentives to improve marketing and quality has been granted in return. European Union funds have been reduced however and the sector which is of significance for Greek agriculture will undoubtedly face considerable hardships in adopting to the new situation. In the wine sector the

Commission proposed a reform mainly oriented toward the reduction of the production potential. Since 1993, all Mediterranean producing countries were encouraged by a noticeable and favorable change in the market situation for the product have resisted the change. Final decisions in the Council were postponed until the situation in the markets is cleared out. Reform has been planned and promoted in the olive oil sector as well, but its thrust still remains unclear. The Commission's main idea is to replace the present support mechanism of production and consumption aids for olive oil by an aid per tree within finite limits.

The tobacco sector has also been under intense investigation and attack by non producer member states such as the United Kingdom while the Commission has been lobbying for a system of voluntary cultivation abandonment by means of financial incentives coupled with an emphasis on the quality aspects and the upgrading of the producer organization role. In order to deal with the beef crisis the Commission designed a series of measures to discourage production. Finally, in the milk sector, no major reform has taken place since quotas were introduced in 1984 - a very important year for dairy policy and the CAP development. Pressure is building up within the sector, from other sectors and international developments. The Commission has been postponing new commitments until prospective future European Union enlargement to the Central and Eastern European Countries become clear and preparations for the World Trade Organization 1999 review are well under way.

The Uruguay Round Agreement on agriculture

Agricultural trade was an integral part of the Uruguay Round negotiations of the General Agreement on Tariffs and Trade (GATT) which began in September 1986 in Punta del Este, Uruguay and extended over eight years to 1994. Agriculture was of no concern during the previous rounds of negotiations. It constituted a case apart as its various features, biological, technical, social, economic and environmental made it fundamentally different from any other sector.

During the negotiations, the European Union sought minimal changes whereas the Untied States advocated a rapid dismantling of all protection. The ultimate agreement had four dimensions, namely: 1. a reduction in the level of domestic support 2. a reduction in the degree of domestic market protection 3. a reduction in the extent of export subsidization and 4. the introduction of sanitary and phytosanitary regulations. The new agreement on agriculture aimed at:

- expanding trade liberalization and increasing competition in international trade;
- minimizing market distortions, by means of the tariffication process. Market signals are better transmitted through the price mechanism and supply adjusts more easily to demand;
- encouraging specialization in production, making more rational use of production and of natural resources;
- promoting the adoption of policies with well targeted measures, well defined framework of operation, and minimum impact on third markets;
- speeding up the structural adjustment process in the agricultural sector;
- setting the conditions for world market stability, by means both of improved access and of reduced trade distortions;
- creating new job opportunities in competitive sectors of the economy in the long run;
- raising world prices in the long run by means of reducing the volume that would be disposed for absorption in the world markets;
- reducing the social and financial cost of policies;
- effecting lower consumer prices; and at
- facilitating the adoption of innovations and technological improvements.

In the short run, the agreement may bring non-competitive agricultural enterprises to a disadvantage. Non-competitive agricultural firms will probably have to confront reductions in income and increases in unemployment. The European Union endorsed the above aims on condition that the principles of the Common Agricultural Policy were not called into question and proposed a gradual but substantial cut in agricultural support based on two principles, the balanced global approach with no separate commitments in the individual commodity areas and the utilization of an aggregate measure of support to be worked out by common accord.

In November 1992 an agreement was reached in Washington, D.C. (Blair House Accord) between the European and the American delegations. In the area of Internal Support, a gradual cut of 20% in support prices until 1999 was agreed upon with a reference period corresponding to the years 1986 to 1988, whereas direct income subsidies were not to be affected. The agreement reached does not make clear what the position will be after 1999 with regard to the direct income subsidies granted under the new CAP.

The Uruguay Round Trade Agreement was implemented in July 1995 under the auspices of the World Trade Organization. The agreement, as it relates to agriculture, extends throughout a six-year period to the year 2001. All parties involved are committed to reopen negotiations in 1999 to reach a new agreement which is expected to be more radical mainly because of changing political circumstances in the European Union. Throughout the Uruguay

Round negotiations the European Union was represented by the Commission. In the Council however, Ministers of Agriculture had the opportunity to express each member state's national position and preoccupations. Greece insisted (Papaconstantinou, 1991) that Mediterranean products such as fruit and vegetables should be excluded from the Agreement on Agriculture by arguing that such products do not contribute to the Community's structural surpluses thus leaving international trade practically unaffected. The argument was rejected by other member states on grounds for the need to strike a balance among all member states with respect to concessions, as well as by other partner states in GATT which forcefully demanded the inclusion in the agreement of all agricultural products with no exceptions.

With respect to domestic support measures the final Act provided for lower support price reductions for products of significance for Greece such as cotton, tobacco, fresh and processed fruit and vegetables, wine and olive oil, as compared to average reductions. Another point of concern to Greece at the time was the effect the outcome of GATT negotiations would have on the CAP financial transfers to the country. It was estimated (Papaconstantinou, 1991) that financial transfers would only be reduced by 6 % for a period of five years, corresponding in absolute terms to ninety billion drachmas which was judged as an acceptable sum, but not high enough to alter the general positive picture of EAGGF expenditures in Greek agriculture. The CAP reform as well as the Uruguay Round Agreement on agriculture attracted strong criticism in Greece from public opinion, mass media, political parties, producer organizations and experts. Farmers expressed their discontent in the form of fierce demonstrations and disruptions to the national transportation network focusing their opposition on issues such as the reduction of prices, reduction in the total value of output, high production costs and income reduction.

The outlook for the future

The major CAP reform, agreed in 1992 and implemented, mainly for internal reasons, in 1993, paved the way for the acceptance by the European Union of a more market oriented approach in agriculture and agricultural trade in the context of the Uruguay Round Agreement. The reform was unavoidable as the substance, mechanisms and instruments of the CAP were subject to ever increasing attacks by third parties, such as the United States and the European Union as faced with radical demands which, if accepted, would have led to the dismantling and an overhaul of the entire policy. The Blair House historic agreement, between the European Union and the United States was considered as being closer to the European position than that of the United States and there was little doubt about the fact that the internal support aspects of the agreement were very satisfactory to Europeans. Especially the extension of compensatory payments from the annual reduction commitments leaves the European Union a large margin of discretion to decide how it wishes to organize and run its internal agricultural and rural policies. With respect to market access, a sufficient level of Community preference was maintained and any additional access opportunities were kept within manageable proportions. The expected rise in world prices was expected to increase the European Union's relative competitiveness. The possibilities of unsubsidized exports were expected to offer increased opportunities to exporters although emphasis had clearly shifted to high quality products and processed food.

For all member states, the Uruguay Round is probably the most important factor that will shape future developments in the international scene. Furthermore, developments taking place in Central and Eastern Europe which are related to the Union's enlargement are at least equally important for Greek agriculture and agricultural trade. In the past, the European Union had been a major beneficiary of this market outlet. Greece, in particular, has benefited by exporting substantial quantities of citrus fruit for example, a fact which partially explains the delay in the adjustment process in the sector with respect to quality improvement and restructuring. Currently, budgetary constraints, the lack of foreign exchange along with the deeply rooted crisis confronting these countries hinder a further increase of such exports from any European member state. In the long term however, given the great potential the countries of Central and Eastern Europe possess, agriculture will play a major role in the reform process provided a stable environment that will allow for long term planning is guaranteed. Developments in Central and Eastern Europe do provide new challenges and opportunities for Greek agriculture although market access will be further increased for products originating in these neighboring countries.

The 1992 CAP reform aims at rendering it more market oriented and seeks to maintain a vibrant rural community by means of direct income support and the implementation of an ambitious rural development program. The Uruguay Round Agreement on Agriculture is compatible with the reform by allowing direct income payments to farmers and rural citizens. Farmers in Greece have a stable policy framework to adjust to by taking advantage of the new opportunities. What is missing is a clear and realistic Community and national strategy for the future adjusted to the specific characteristics of each part of the Greek countryside where agriculture is one activity among others. In agriculture, paying farmers the low prices set at world market levels which bear no relation to the country's agricultural structures particularly that of small farmers and farmers in the marginal mountainous and remote coastal areas would lead to economic losses. When social assistance in the form of income compensation is calculated on a flat rate basis which discourages individual effort and hinders modernization, economic losses will certainly not be offset. Unless a large number of agricultural holdings disappears in Greece, the implementation of the new Common Agricultural Policy will not ensure a permanent solution to the problems of correcting regional imbalances and inequality of agricultural incomes, since the new system gives preference to areas with higher yields.

Estimates of producer gains and losses, as a consequence of the CAP reform as proposed by Commissioner MacSharry (European Commission, 1991) are provided by A. Sarris (1994). The net effect of the proposed arrangements for crops and livestock would be to continue the previous high level of support for all but the larger EU farmers based on historic criteria of output and income while preventing it from increasing. For the case of Greece, the concept introduced by MacSharry would lead to a net benefit from cereals, milk and beef farms of any size, with benefits per farm being much greater for larger farms (table 6.1). Significant losses would occur in the sheep meat and tobacco sectors for almost all size classes and especially for average size farms due to the fact that no income compensation was envisaged for tobacco and sheep meat producers. In general, loss per hectare is much larger for smaller farms and it is only the category of very large farms, (over 100 hectares) that stand to benefit from the MacSharry reform. The average loss in Greece was estimated at 68.3 ECUs per farm and the total loss of producers' benefits at 65.1 million ECUs per year. Of the total losses to farmers in Greece, 80% would accrue to small agricultural holdings of sizes less than 5 hectares which account roughly for three quarters of all farms.

The MacSharry reform proposals were altered by the Council of Ministers which increased the proposed payments to ensure that most of the larger farms also receive full compensation. This change practically leaves large farms in Greece unaffected as they are not large by any North European standards with the exception of the largest class which refers to a negligible number of large farms (nine hundred farms). It should be noted that the level of compensatory payments was calculated in relation to the fall in support (institutional) prices rather than market prices. Therefore, if market prices' fluctuations do not coincide with the reduction in support levels, a development that can very well be observed for demand and supply or currency realignment reasons, then a large number of farms in some member states and particularly large farms outside Greece can experience a rise in market prices along with full scale compensation and a substantial rise in incomes. This was the case at least for the United Kingdom for the 1995/1996 marketing year (Sheehy, 1997). By the year 2001 when the next World Trade Organization (WTO) Round is scheduled to begin, protection reduction will have progressed enough to

Table 6.1

Impact of CAP reform on agricultural holdings in Greece by size [*]

Farm size (ha)	<1	1to<2	2to<5	5to<10	10to<20	20to<30	30to<50	50to<100	>100	All
Number of farms (thousands)	249.8	191.8	296.2	140.7	53.5	11.3	6.2	2.9	0.9	953.3
Average size (ha)	0.48	1.38	3.14	6.74	13.29	23.52	36.86	62.98	213.53	4.03
					Net impact per holding (ECU)					
Cereals	0.1	0.3	1.0	2.8	6.3	11.2	17.0	28.3	35.5	1.5
Milk	1.2	3.3	7.5	14.9	19.2	26.0	23.0	38.8	76.1	7.3
Beef	0.5	0.8	1.3	2.5	3.7	11.8	5.8	6.7	13.3	1.5
Sheepmeat	-5.5	-7.0	-13.3	-23.9	-48.6	-87.2	-125.5	15.9	0.0	0.0
Tobacco	-22.5	-66.2	-105.4	-115.0	-124.3	-135.9	-120.1	-80.4	-88.9	-78.6
Total benefit per holding (ECU)	-25.9	-68.7	-108.8	-118.7	-143.7	-174.1	-199.7	-22.5	36.0	-68.3
Total benefit per hectare (ECU)	-53.5	-49.7	-34.6	-17.6	-10.8	-7.4	-5.4	-0.4	0.2	-16.9

Source: Sarris, A., 1994

---

[*] Calculations are based on data from the 1987 farm structure survey of the National Statistical Service of Greece

affect prices, incomes and trade. External pressure for further liberalization is expected to increase coupled with additional European Union concerns about the expected enlargement that will include countries of Central and Eastern Europe. The Common Agricultural Policy is therefore expected to undergo a further reform that ought to be designed in such a way as to satisfy WTO partners as well as to facilitate enlargement. It is very likely that such an attempt will begin soon before the completion of the next trade negotiations. The most probable scenario of a new CAP reform is that of "further developing the 1992 approach" (European Commission, 1995). Price levels are expected to be brought further down toward the world trading level, and compensation by direct payments will most likely be provided where necessary. Such compensation however will probably be linked to environmental and social considerations in the framework of a region specific integrated rural policy aiming at a more balanced geographical spread of economic activity.

Producers in Greece who will probably suffer more from adjusting to the new international policy conditions will be those who have traditionally concentrated their production in heavily supported products irrespective of efficiency considerations and quality. As a consequence to the international policy reform toward increased market orientation the unit revenue will drop and farmers will cease agricultural activity. With reduced domestic intervention and less distorting effects in the world markets, relative prices will change, thus influencing both the patterns of production and trade. Production patterns will be affected more by factors determining the holdings' relative costs, the country's comparative advantage and the locations' specific characteristics and potential. Changes in the value added as well as in the traded quantities will affect the national trade balance and consequently foreign exchange revenues. Equally important, under increased market orientation, consumer preferences will assume a new significant role in determining price levels for each produce, affecting the final product mix and production techniques. Greek agriculture currently has no significant potential to take better advantage of economies of size. Alternative strategies for sustainable agriculture should be investigated and designed for genuinely sustainable rural development possibilities that would prevent the out migration of young people from rural areas, land abandonment and environmental degradation.

The Uruguay Round Final Act as well as the CAP reform will serve as catalysts in forcing the sector to face up to the circumstances of increased competition in the world markets by bringing about substantial changes in Greek agriculture. In conventional agriculture, this translates into further mechanization, adoption of capital intensive techniques, higher unemployment in rural areas and a more skewed distribution of income.

177

Greece must find ways to adapt to the upcoming transition as smoothly as possible. New strategies should be developed in order to accommodate the problems facing the large and small farmers alike.

Following the conclusion of the Uruguay Round multilateral trade negotiations, several attempts have been made by independent researchers as well as international organizations to assess the impact of the agreement on agriculture. The main conclusions of such analyses, at a global scale, may be outlined as follows: agricultural trade will be limited (Tangermann, 1994, Delorme and Clerc, 1994). Tariffication is not expected to exert a significant influence on trade flows as well as on agricultural product prices in the coming years (Ingco, 1995; Hathaway and Ingco, 1995). Commitments on minimum access and domestic support are not expected to exert any significant influence either. Direct effects are expected only from commitments on subsidized exports and especially commitments on volume reduction (table 6.2). International prices are generally expected to increase as compared to prices which would have prevailed without the agreements. No substantial influence on the volume of international trade is expected although trade flows and production patterns may change as Western Europe and Japan are expected to increase their net imports of basic agricultural products while North America and Oceania are expected to have export gains (FAO, 1995; OECD, 1995).

From a social welfare point of view, the implementation of the Uruguay Round Agreement on agriculture is expected to create significant benefits at the international level. Such benefits are expected to accrue to the developed world, namely, the United States, the European Union and Japan. Developing countries and especially net food importers are expected to have losses due to the phasing out of subsidies in North America and Europe. In sum, although the degree of liberalization is judged as moderate the agreement is considered a very significant one as it contains elements with important and permanent agricultural policy consequences.

Market access

In the market access domain the Agreement on Agriculture provides that all non tariff barriers have to be converted into ordinary customs duties in the form of tariff equivalents based on the difference between average 1986-1988 (base period) domestic prices and international prices. Ordinary tariffs and tariff equivalents have to be reduced by 36% with a minimum of 15% per tariff line for any agricultural commodity from the base period level. Reductions should be implemented in equal installments during the period of implementation. For tariffied commodities, Greece as an EU member is obliged to maintain existing market access conditions including existing preferential arrangements granted to specific exporting countries if they were already in excess of minimum access commitments during the base period. For all tariffied commodities whose imports were below a minimum level in the base period, Greece is obliged to establish minimum access opportunities representing 3% of the base period domestic consumption in 1995 (the first year of implementation) rising to 5% in the year 2000. Additional duties may be imposed in the case of import surge or of low prices based on the average 1986-1988 reference price for the product concerned.

Table 6.2
Reduction commitments on subsidized exports
(EU12, volume in '000 tons, expenditures in million ECUs)

Commodity	Base [a]	1997	1998	1999	2000
Wheat					
Volume	17,008	16,846	15,709	14,573	13,436
Expenditure	1,783	1,698	1,512	1,327	1,141
Olive oil					
Volume	148	133	127	122	117
Expenditure	86	70	65	60	55
Fresh fruit and vegetables					
Volume	1,148	1,027	987	947	907
Expenditure	102	84	78	72	66
Processed fruit and vegetables					
Volume	201	180	173	166	159
Expenditure	15	13	12	11	10
Wine [b]					
Volume	3,080	2,757	2,649	2,541	2,434
Expenditure	65	53	49	45	41
Tobacco					
Volume	143	159	144	128	113
Expenditure	63	73	62	51	40

[a] Base reflects level in 1986-1990
[b] Quantities in 100 liters

Source: A. Sarris, G. Mergos and P. Sarros, 1996

A recent study published by the Foundation of Economic and Industrial Studies in Athens (Sarris, Mergos and Sarros, 1996) estimates, by means of quantitative modeling, the impact of the Uruguay Round Agreement on Greek agriculture. With respect to the market access provisions of the Uruguay

Round Agreement on agriculture various issues concerning commodities of special interest to Greece are extensively treated in the study. The most important of such findings are briefly outlined.

In the cereals sector, tariffs to be imposed through the year 2000, according to the agreement, are expected to be higher than the tariffs applied in 1995. The obligation of the European Union however to establish a minimum access opportunity for 300,000 tons of wheat might impede Greek exports of durum wheat, a commodity of importance for the national agricultural economy, to the rest of the Community during periods of excess supply. On the contrary, the tariff equivalent for imports from Mediterranean countries which are not members of the European Union is expected to increase further according to the agreement due to the low product transportation costs as compared to exports from the United States or Canada. Thus, imports for durum wheat from neighboring countries, such as Turkey which is the major potential exporter in the area for this commodity, will be made more expensive. No other major changes are expected to affect the country in the cereals sector.

In the olive oil sector, the European Union has agreed to lower tariffs for imports from Lebanon, Turkey, Tunisia and Morocco. Olive oil produced in Greece as well as in the other Mediterranean member states, mainly Italy and Spain, will continue to dominate the European markets whereas third country imports will be observed only during periods of reduced supplies.

In the wine sector quantities expected to be imported into the European Union and thus, Greece, do not seem to lead to an intensification of already existing problems. Market instability in the Union is due to the excess production of low quality table wine. It is the common organization of the markets reform for wine and not any possible but moderate increase in imports, of the agreement that will influence developments in the sector, in spite of the expected reduction of tariffs. Imports of third country origin wines into Greece are not expected to increase. Neighboring wine producing countries such as Bulgaria are not expected to increase their exports to Greece, as tariffs are expected to remain high. Italian and French wines will continue to be competitive in the domestic market due to their popularity among Greek consumers as well as their low prices.

For fruit and vegetables, a reduction of protection against imports has been agreed upon. As far as tariffication is concerned, the European Union, contrary to the provision of the Agreement has declared its firm intention to maintain its own system of reference price, import price and countervailing duties. Thus, a maximum equivalent duty, expected to be reduced by 20% until the end of the period (year 2000) will continue to be imposed upon imports for tomatoes, cucumbers, artichokes, citrus fruits, table grapes, apples, pears, apricots, cherries, peaches and plums. This protection measure, to be applied throughout the year as opposed to a seasonal basis, is comparable to the level of effective protection the European Union has been experiencing during recent years. Excess concessions were offered by the Union at the very late stages of the negotiations for certain products such as almonds and walnuts as a result of pressure exercised by major exporting countries such as the United States. For these products it was agreed that import duties will be decreased by 50%. Such developments do not constitute any major threat to domestic production. No major changes in effective protection are expected through the year 2000 in the sector of oil seeds and protein products, rice, sugar, cotton, tobacco, beef, pork and poultry meat, milk and milk products, processed fruit and vegetables, dried raisins and orange juice.

Export competition

In the areas of export competition the final act of the Uruguay Round Agreement provides for the prohibition of export subsidies for products not specified in any country's schedule. All existing direct export subsidies, the disposal for export of public stocks at subsidized prices, subsidizing the marketing or transport costs of exportable commodities are all subject to a reduction. Reduction commitments refer to both expenditures as well as subsidized quantities. Total expenditures for export subsidies are to be reduced during the period of implementation, from 1995/1996 to 2000/2001 by 36% from their base period level (1986-1990). Subsidized quantities are to be reduced by 21% in a similar manner and final levels with respect to expenditures and subsidized quantities should reflect full compliance.

The same study (Sarris, Mergos and Sarros, 1996) highlights a number of issues regarding export competition provisions for the case of Greece. The study points out that in the cereals sector and more specifically for durum wheat a possible overrun by the producer member states, i.e. Italy, Spain, France and Greece will cause a reduction of the Community support price. The maximum quantity eligible for support at the Community level is 85 million tons. Such a development will mostly affect Greece as it is the country which still has the possibility to substantially increase its total agricultural area under durum wheat cultivation and thus increase production. Another problem pointed out in the study concerns the exportation of oranges from Greece to countries of Central and Eastern Europe. A reduction of subsidized quantities exported is expected as a result of the relevant provisions of the agreements. In the marketing year 1996/1997 such a reduction did occur.

In the beef meat sector the study concludes that due to the reduction of subsidized quantities, commercial activity will shift toward intracommunity trade. Such a development will have a negative impact on the domestic market in Greece because imported beef meat will be offered at lower prices in the domestic market causing a reduction of domestic producers' prices. A similar development is expected in the dairy sector in which a further reduction of prices of imported cheese into Greece is expected for the same reasons.

Domestic support

The area of domestic support contains concessions relevant to the use and the level of non-border measures implemented to support agricultural production. The main provisions of this agreement are summarized as follows: measures affecting production, consumption and trade (Red Box Measures) such as market price support and other payments are summarized in an indicator identified as aggregate measure of support. The total aggregate measure of support for all products is to be gradually reduced by 20% during the implementation period (1995-2000). Any direct payments relating to production limiting programs (Blue Box Measures) are exempted from the aggregate measure of support and are not subject to reduction. Any non or minimally trade distorting measure (Green Box Measures) such as public stockholding for food security purposes, domestic food aid, direct payments to producers that are not dependent on the type or volume of production, payments for relief from natural disasters, producer retirement programs, investment aids, payments under environmental or regional assistance programs are not subject to reduction commitments.

The market liberalization process in Central Eastern Europe, the Balkan and the Black Sea Region (CEECs)[3]

Following the collapse of the communist regimes in Eastern Europe, the CEECs took the path of transition toward market economy. Despite the differences in the pace, the smoothness and the degree of the recorded adjustments, the CEEC agricultural sectors experienced contraction of production and total demand, disruption of the input and output supply systems, deinvestment, increase in unemployment and reorientation of trade.

The weakness of the institutional and legal framework to settle property rights constituted a serious impediment in the process of land privatization and became a major factor of instability and disorganization in agricultural production. Unsettled property rights caused insecurity, shortage of credit, and pushed interest rates in agricultural credit at very high levels. Privatization caused turmoil, especially in sectors where production of output or services was concentrated in large state or collective units. Thus, livestock production, the input and output distribution network and the food processing industry were particularly hit.

182

The farm holdings which emerged from the privatization process were too small to be efficient, while farm investment shrank (Nallet and Van Stolk, 1994). Productivity dropped, labor exited the agricultural sector and unemployment rose since other sectors were unable to absorb the surplus.

In the food processing and the farm input supply industries privatization moved very slowly. Furthermore, it resulted in disruptions in supply and caused decreases in production, widening divergence between farm gate and retail prices, deterioration of rural and urban terms of trade, and drop in demand (Tarditi et al., 1994; Buckwell et al., 1994). The outcome of correcting the price distortions of central planning was similar. Uncertainty prevailed as the CEECs moved from central planning to a liberal regime (1991) and later retreated to protectionism. In addition, instability in macroeconomic conditions added to the uncertainty and eroded attempts for agricultural development (Buckwell et al., 1994).

On the human consumption side, reduced or eliminated consumer subsidies and inefficiencies in the downstream sector caused retail prices to rise. Moreover, the food processing industry, making use of dated technology, could not escape the high cost of credit and of producing high quality food products. As consumer's purchasing power decreased and inequalities in income distribution expanded, domestic human demand contracted and high-priced products (e.g., meat and dairy products) were substituted by food products of lower price (e.g., cereals) in consumer's bundle of goods.

In addition to the fall in domestic demand, the CEECs lost the markets of the former USSR. Nevertheless, a dramatic increase in trade with the European Union (EU) partially compensated for this loss and turned the EU into the CEEC's most important trade partner, accounting for over half of most CEEC's agri-food imports and absorbing about one third of CEEC's exports. EU exports became attractive as CEEC consumers familiarized themselves with Western standards and CEEC exports rose by exploiting the comparative advantage ensued from lower labor costs. Yet, EU imports of CEEC products increased at a lower rate than EU exports, mainly because of differences in quality and sanitary standards. In general, the EU trade balance with the CEECs improved at the expense of the CEECs. Nevertheless, the share of transactions with the CEECs in total EU trade remained at low levels.

CEEC trade was also obstructed by: inflation and currency non-convertibility; weak institutional and physical infrastructure; the inexperience of CEEC traders; and direct trade control, especially through export licensing and quotas. On the other hand, geographical position assisted in developing trade. In the case of Greece, trade with Bulgaria expanded dramatically, and trade with FYROM and Albania showed promising signs, whereas the Greek share in other CEEC markets began to shrink (Wallden, 1995).

Recent developments of income growth in most CEECs suggest that there is a potential for increases in the demand for food and for a rise in imports. This potential is of particular interest to Greece, since the country shares a similar background with several CEECs regarding common history, culture, and religion embodied in the geopolitical position. Greece also enjoys advantages by specializing in products of kind and quality currently in high demand in the CEECs. Moreover, Greek enterprises are more familiar with the operation of a firm having the size and the mentality existing in the CEECs and are more flexible in handling and overcoming poorly organized trade and marketing functions. This will not be an advantage for long. Greece therefore must become more competitive in order to maintain or to increase its market share in the CEEC or the EU markets.

The realization of the potential for development and economic cooperation depends on the establishment of a stable environment and on socio-economic cohesion. These objectives can be best served by simultaneously promoting: 1. the development of EU-CEECs relations; 2. the adoption of appropriate agricultural policies by the CEECs; and 3. the development and strengthening of relations between Greece and the CEECs.

The European Union's interest in enhancing socio-economic cohesion and development can be channeled through facilitating the CEEC's adjustment to the market economy, the operation of the markets, CEEC's accession to the EU, as well as through promoting the creation of jobs, improvements in competitiveness, environmental protection, cross-border and regional cooperation and equal opportunities.

Since smooth accession is essential for stabilization, security and economic strengthening in the entire continent, it is imperative that European integration be well prepared and proceed at the lowest possible social cost of adjustment. Besides the financial forms of assistance, the EU should help the CEECs to develop and to harmonize their agricultural sectors with EU agriculture.

The CEECs accession to the EU is expected to increase total welfare as a consequence of the elimination of border measures and of the increase in trade flows. Nevertheless, the impact on agricultural trade will depend on the future of the CAP, on the conditions governing the accession, and on the specification of comparative advantage under the new conditions.

On the other hand, the CEECs should aim at developing an efficient and competitive agricultural sector, by securing free entry into the markets. Thus, privatization of land, processing firms, the financial sector and reduction of barriers to entry should be promoted simultaneously. Moreover, the inefficiency of the small farm holdings should be weighed against the benefits accruing from their role as absorbers of surplus labor (Buckwell et al., 1994).

In order to develop a credible farm credit system, property rights should be clearly specified and a land market should be established. Moreover, it is imperative that production in the CEECs should not fluctuate widely. Producer prices should be stabilized at a level covering the average variable cost of production. Nevertheless, it is important that a system of price supports minimize market and trade distortions. Such a policy reform, in combination with the international tendency toward market deregulation and trade liberalization, could result in major changes in the patterns of trade.

The CEECs should also adjust their exchange rates, promote monetary and price stabilization, agricultural research, training and extension services. The quality of their products must be improved, free entry in the food industry has to be secured, economic and technical ties should be prepared and developed for gradual policy harmonization.

Finally, Greece should pursue the strengthening of European policies and institutions, a balanced EU foreign policy, and the parallel promotion to the EU of structural policies for socio-economic cohesion and solidarity, of the CAP reform, and of institutional changes to accommodate the enlargement. In doing so, however, Greece should safeguard the geopolitical balance between regions by promoting the accession of Cyprus and Malta, as well as by defending Russia's security interests and the establishment of a long term mutually beneficial cooperation between the EU and Russia (Wallden, 1995).

The EU enlargement to the CEECs and the need for reform

The development and the strengthening of the cooperation with the CEECs will benefit these countries, the EU as a whole, and Greece in particular. Political and socio-economic stability and security are prerequisites for economic development in the CEECs. On the other hand, a stronger Europe in the international scene implies prosperity for all European countries. Promising development of the EU economy, however, requires the enhancement of social and economic cohesion, which necessitates the diminution of regional differences in development. Besides, stability influences the rate of migration to the Mediterranean region, which affects the development of Greece. Thus the shared interest for stability and prosperity constitutes the main motive for promoting wider cooperation and the accession of the CEECs into the EU.

Another factor urging the strengthening of cooperation is the growing interdependence in the world and within the EU, ensued by trade liberalization and European integration. As a consequence of the international division of labor, creation of jobs should be dealt with within a wider macroeconomic context that integrates employment policies into economic and social policies.

Thus, in addition to the goal of economic growth, it is necessary that a broad approach be adopted by both current EU members and the CEECs, aiming at combating social exclusion and unemployment.

In particular Greece, being the sole Balkan EU member state, has particular interest in promoting not only regional cooperation, but also the accession of the CEECs and the Balkan states in the EU. Such an EU enlargement would imply political benefits for Greece, as well as potential economic benefits in the long run. If differences in development and in socio-economic measures are not assuaged, there will be no cohesion in foreign policy and in security matters. Eventually this could lead to the permanent marginalization of the Balkan region.

Despite the economic costs that the CEEC's accession can imply for Greece in the short run, the country's growth depends heavily on the creation of a unified economic space that offers the opportunity for wide economic cooperation, increase of market share and development. Long run benefits will accrue from a good performance in a competitive international environment and from a dynamic penetration of Greece into the EU, the CEECs and third country markets. Thus, despite the economic superiority that Greece could enjoy over the CEECs for years to come, it is necessary that the Greek economy improve its competitiveness (Wallden, 1995). In addition, Greece's long-term, political and economic interests depend heavily on the role that it will undertake toward the accession of the CEECs and especially of the Balkan states into the EU (Wallden, 1994). It should not be underestimated that economic cooperation especially with the Balkan states would not only have a positive impact on agricultural economy, but also would prevent regions near the border from being depopulated.

In view of the future accession of the associated countries of Central and Eastern Europe, the issue of alternative strategies for agricultural and rural development is gaining in importance. Enlargement is expected to almost double the current EU agricultural labor force of over 6 million persons, to increase agricultural land by 50% and to add over a 100 million food consumers of very low purchasing power, relative to the currently prevailing EU level. Price level alignments are expected to raise cereal, beef and milk surpluses, thus increasing budgetary costs at politically unacceptable levels (CEC, 1995).

In the light of such developments, a major reform in the CAP as well as in the operation of the Structural Funds will be unavoidable. Nevertheless, a radical reform that would adopt an across the board abolition of support prices, elimination of supply management measures and phasing out of Community funded, direct income compensation would be an undesirable development for Greece, the CEECs, and other EU member countries. Such a scenario would imply the renationalization of agricultural policy and would

186

require huge sums from national budgets for supplementing farm income. In this case, the gap between poorer and richer member countries and regions would widen, shaking the validity of social cohesion. Desirable policy reforms for Greece and the CEECs would be in the form of further progress toward: 1. improved competitiveness and greater participation in world market developments and 2. promoting an integrated, multi-functional rural policy.

In particular, targeted and direct farmers' income compensation, better integration of market policies and sustainable rural development would be best suited for achieving better geographical balance of economic activity and development in Greece and the CEECs. The reformed common agricultural, structural and rural development policy should take into account the diversity characterizing the agricultural sector and rural areas across the EU and the associated CEECs. National and regional authorities should be provided with funds sufficient for the implementation of long term framework programs aiming at structural improvement in agriculture and integrated rural development. Furthermore, it is necessary that the specific physical, economic, social and institutional characteristics be addressed.

On July 16, 1997, the European Commission presented to the European Parliament its detailed strategy for strengthening the European Union and preparing it for its eastward enlargement, entitled "Agenda 2000". The strengthening will be based on four main axes: 1. further institutional reform and review of the Commission's organization and operations; 2. the development of internal policies for growth, employment and quality of life; 3. the maintenance of socio-economic cohesion through more effective Structural Funds; and 4. further reform of the Common Agricultural Policy (CAP).

It is interesting to note that the Commission acknowledged the importance of cohesion and took into account the increase in income divergence that will result from enlargement.[4] It decided to reform the Structural Funds, so as to have more targeted and effective function and to promote sustainable rural development.

Thus, the current seven objectives for regional programs are being reduced to three, covering the new Objective 1 - or lagging in development - regions (previously falling under Objective 1 and 6), the new Objective 2 - or declining industrial (previously Objective 2) and rural (previously Objective 5a and 5b) - areas, and the new Objective 3 regions - or the regions in need of adapting to the new labor market conditions and the need for changes in the industrial sector (previously Objectives 3 and 4). The corresponding programs will be promoting mainly competitiveness, economic diversification and modernization of the education, training and employment systems respectively.

In addition to the reorganization of the Structural Funds, the Commission proposes "a more prominent role to agri-environmental measures, especially those which call for an extra effort by farmers, such as organic farming, maintenance of semi-natural habitats" (CEC, 1997) in order to pursue a more effective rural policy.

Although stricter criteria would apply for eligibility,[5] it is proposed that the new Objective 1 regions would be allocated about two thirds of the Structural Funds available for the 15 member states and that the Cohesion Fund will remain unchanged.

On the other hand, the final proposals for the arrangements covering tobacco, olive oil, and wine are still pending while the proposed CAP reform affects mainly cereals, beef, and milk. These crop and livestock subsectors would be facing price cuts of 20% in 2000, 30% between 2000 and 2002, and 10% by 2006 respectively, which would be compensated by direct (i.e., per hectare, per head) payments.

The Commission proposal presents an opportunity for Greece to take advantage of the favorable reform and to negotiate for further improvements in the final arrangement. During the process of evaluating the impact of "Agenda 2000" on Greek agriculture, it should be kept in mind that Greece's development relies mainly on a strong, and self-sustained rural sector, within which agriculture has a substantial share. Nevertheless, it would be wise to give weight to the impact of the proposed changes, by looking at their relative importance and by setting priorities. Given the importance of sustainable rural development, especially to Greece, it should be reminded that the proposed price cuts refer to products of no particular interest to Greece. As a matter of fact, it is likely that these reductions would benefit Greece overall, since consumers will be able to purchase these largely imported products at lower prices.

Perspectives for Greek agriculture

Needs, problems and constraints

The exposure via trade of a country's structural advantages and disadvantages to an expanded world market makes that country's economy more sensitive to international developments and more vulnerable to foreign competition. As trade liberalization facilitates the transmission of price signals between world and domestic markets, the increasing international competition threatens the less competitive domestic sectors with job losses, income reduction and deterioration of quality of life. On the other hand, globalization makes the access to new markets easier and offers substantial opportunities for gains.

The upcoming WTO[6] negotiations and the expected - in view of the EU enlargement to the CEECs - CAP reform delineate a new international economic environment. The Greek economy can either adjust to this new framework of world market trends or reduce itself to being a marginalized partner. The country's defiance of marginalization, however, requires that the problems hampering socio-economic adjustment be resolved. Therefore not only structural but also institutional inefficiencies should be removed.

In particular, the handicapped agricultural sector would have to confront the challenges by structural and institutional constraints from both economic globalization and the CAP operation. Nevertheless, these challenges should not be confronted in an isolated and sectoral manner, since the allocation of available resources both between and within sectors determines the performance of the economy. Only an integrated and comprehensive strategy could be effective in achieving adjustments. Any attempts to deal with the problems separately would either prove futile or would lead to further distortions.

The problems that stemmed from the CAP operation differ in nature. While the pre-reform period gave rise to inefficiencies, the recently established stabilization of institutional prices effected lower guaranteed farm income levels. The end of increasing price levels emerged - at least in part - as a consequence of the trade liberalization process promoted by GATT's Uruguay Round and by the WTO.

The orientation that the pre-reform CAP intensified production that led to widespread misallocation of resources[7] and to increased income disparities within the rural and the agricultural sectors in Greece. Meanwhile, the induced irrational use of consumable inputs precipitated environmental degradation and depletion of natural resources, especially of water. Furthermore, the CAP altered the terms of trade in agriculture. Thus, it effected changes in relative incentives and in the composition of agricultural production. Given the structural and institutional constraints presented below, this development induced increased production of low value, low income elasticity agricultural products. Moreover, the combination of price supports and of restricted access by producers to full and accurate information brought about institutional changes that were reflected in the adoption of a new mentality by farmers and farm organizations alike. The institutional framework that resulted, acted as a strong inhibitor to restructuring.

The presence of structural impediments, encapsulated mainly in the low mobility of land, labor, and production quotas, obstructed further the attempts to adjust. The prevalent institutions aggravated the difficulties imposed upon by natural constraints.

Despite the increase in rented areas, agricultural land exhibited low mobility, partly because non-agricultural uses began to compete intensively

189

for it from 1970 onward (Sapounas, 1991). To a large extent, agricultural land became a financial asset. High rents and land values, as well as a quota system discouraged or even prohibited the establishment of young farmers. Furthermore, the exodus of labor from agriculture and the early retirement program did not suffice in raising farm sizes adequately.

The mobility of labor, following a period of high movement in the 1960s, abated in the 1970s and was minimized in the 1980s. This resulted partly from the inability of the secondary sector to absorb labor surplus and from the convergence of wages in various labor sub-markets to the equilibrium level. Furthermore, indications suggested that the migratory movement changed direction in the 1980s, taking place from urban to semi-urban and rural areas (Sapounas, 1991). This allegation could be supported by the increased unemployment rates in urban centers and by the policy measures that were introduced in order to provide incentives for returning to rural areas (Sapounas, 1991).

Finally, the apparent high mobility of capital in the agricultural sector during the period 1950-1989 was due to the necessary replacement of obsolete machinery, which, in addition, was favored by the relatively low cost of capital for agricultural purposes. According to Lasaridis et al., (1994), not only did a sizeable part of investment represent machinery (on average 45% and in several regions exceeding 50% of the total value of investment), but also this form of investment was frequently underutilized or operated for non intended uses. Indicative of the validity of this allegation is the fact that in 1989 each holding owned on average 1,02 tractors (Sapounas and Miliakos, 1996), while the average farm holding was smaller than 4.5 hectares. Mechanization took place irrationally and neglected the conditions prevailing in Greek agriculture. Thus, despite the need for low power tractors, farmers invested in the heavily subsidized purchase of high power tractors which were intended for use in large areas of land and for substitution of labor (Mergos and Psaltopoulos, 1996). It follows that such irrational investments failed to improve productivity. Furthermore, investment initiatives usually entailed some form of subsidization or financing, since the value of production was insufficient for such an undertaking (Sapounas, 1991).

The rigidities in the agricultural sector perpetuate in a setting that is characterized mainly by the following (Sapounas and Miliakos, 1996; Labropoulou, 1995; Lasaridis et al., 1994; Sapounas, 1991):

- the ageing population of farmers (in 1988, 63.6% of the agricultural labor was older than 48 years of age) and especially of farm managers, almost 60% of whom are older than 56 years of age (Sapounas and Miliakos, 1996; Labropoulou, 1995);

- large percentage of farm holdings (70%) are in mountainous and disadvantaged areas (Labropoulou, 1995);
- extended dry seasons and, in general, adverse climatic conditions;
- seasonal production;
- Specialization in single activities and in products volatile to weather conditions. This characteristic is accountable for large fluctuations in production and income levels;
- low investment levels, incomplete land reclamation, low percentage of irrigated land (33.8% of total arable land in 1991) and poor irrigation systems (Labropoulou, 1995);
- farm structure of small sized and fragmented family-operated holdings;
- inefficient and dated operating institutions and organizations, such as the Agricultural Bank of Greece, public organizations, and cooperatives;
- absence of participation by farmers in the operation of institutions, including agricultural cooperatives;
- weak labor market structure in rural areas;
- inadequate incentives in rural - and especially in mountainous, less favored, and remote - areas for preventing the population from migrating;
- poor infrastructure in the sectors of marketing (storage, conservation, processing, grading, packaging, transportation, distribution);
- weak information systems and inadequate technical support to producers;
- poorly organized marketing and trading systems domestically and abroad;
- lack of quality control mechanisms;
- lack of transparency in transactions;
- very poor education and training level of farm labor and of farm managers;
- absence of agricultural extension services, lack of research and of biotechnology applications; and
- high cost of financing investment.

More specifically, the agricultural sector has been operating as a residual sector in the economy, sustaining mainly low skilled labor with low opportunity cost and low likelihood of being absorbed by other economic sectors. Thus, despite the migratory movements, agriculture still constitutes an important employer in Greece and assumes an important social role depicted in the family character of nearly all farms. Nevertheless, among the people comprising the farm labor force, only one in every ten persons is occupied exclusively (on a full-time basis) in on-farm activities (table 6.3).

Table 6.3
Farm labor force, 1993

	Total farm labor as % of the EU12	Full-time as % of total farm labor force	Family labor as % of total farm labor
Greece	11.8	9.7	99.7
Spain	17.1	21.3	94.5
France	10.7	39.1	89.6
Italy	31.7	12.4	98.5
Portugal	8.4	11.5	94.8
EU12	100.0	22.2	93.6

Source: Calculated from data presented in *Statistics in Focus,* 1995, Eurostat

Damianos et al., (1994) in an empirical study about the role of multi-activity in the Greek agricultural sector concluded that 42% of farm managers and 50% of family labor maintain off-farm lucrative activities, that is, they engage in multi-activity. In addition, it was shown that the flexibility exhibited by farm labor in seeking other sources of income increases with the level of skills and education and decreases with the levels of wealth and net value added in agriculture. As a consequence of widespread multi-activity, agricultural activities contribute barely two thirds of the total income of total farm holdings and less than 25% of the total income earned by multi-active family holdings.[8]

This extended phenomenon of part-time farming should draw the attention of policy makers to the invaluable contribution of off-farm employment in the formation of total farm income, especially for small farm holdings. Policies aiming at maintaining a fair standard of living for small farms should focus on establishing a nourishing environment for alternative employment opportunities rather than on enforcing price support mechanisms that favor large scale operators, thus widening disparities and inequities.

Another factor constraining the development of the agricultural sector, ceteris paribus, is the age structure of the farm managers (table 6.4). In Greece about 60% of the farm managers are over 55 years of age. In addition, almost 60% of the farmers who are older than 55 years of age are found on small holdings.

Sarris (1994) addresses another structural obstacle by arguing that "the problems of southern EC agriculture are not of inadequate income per farm or inadequate capital per farm, but simply of larger number of small farms relative to the north. This might be due to land constraints coupled with the history of land-tenure systems, as well as to the overall lower level of development and lower overall level of capital". Merlo and Manente (1994) also claim that the determinant of low labor productivity and income levels is the small average farm area, which mainly is attributed to natural factors, such as the climate and the high percentage of mountainous areas.

Table 6.4
Age of the manager, 1993 (%)

	< 35 years	35-54 years	55-64 years	> 65 years	Total
Greece	7	36	28	29	100
Spain	8	39	31	22	100
France	13	45	27	15	100
Italy	6	33	29	33	100
Portugal	5	33	29	33	100
EU-12	9	38	28	25	100

Source: Calculated from data presented in *Statistics in Focus,* 1995, Eurostat

The problem of high production costs facing Greek agriculture stems partly from the fact that more than three-quarters of all holdings are smaller than 5 hectares in size, whereas the percentage of holdings in the size classes exceeding 49 hectares is almost negligible (table 6.5). Taking into account Hill's (1984) claims that an agricultural enterprise needs 150 to 200 hectares in order to take full advantage of the "available economies of size which modern production methods entail", it follows that conventional agriculture in the European Mediterranean region is disadvantaged in competing with other European and third countries. Stated otherwise by Larsen and Hansen (1994), "while the major gains generally are realized for medium-sized and large farms, the data indicate that economies of size might not be fully realized for most countries by the largest farm types represented". Data from the 1980s for the ratio of total costs to total production value, show that the cost share in Greece and the other Mediterranean European countries was minimized for farms greater than 44 hectares in size.

The above mentioned constraints contributed to the realization of low investment levels. Consequently the insufficient use of capital was depicted in high labor to capital ratios and implied the use of relatively less capital intensive techniques. In the context of conventional agriculture, this result accounted for weak productivity performance, when measured in real net value added per annual work unit.

Table 6.5
Distribution of agricultural holdings by size class
of agricultural area (AA), 1993

Farm size	< 5 ha (%)	5-20 ha (%)	20-50 ha (%)	50-100 ha (%)	> 100 ha (%)	AA by holding (ha)
Greece	76	21	2	0	0	4.3
Spain	58	27	8	4	3	17.9
France	28	23	26	17	8	35.1
Italy	77	17	4	1	1	5.9
Portugal	78	17	3	1	1	8.1
EU12	59	23	11	5	3	16.4

Source: Calculated from data presented in *Statistics in Focus,* 1995, Eurostat

Another problem stems from the unbalanced composition of agricultural production in favor of low value added crop products. The composition of Greek agricultural production differs considerably from that of the Northern EU member states. Greece is mainly producer of the so-called Mediterranean, either crop or livestock products, which are moderately supported by the CAP.

Fruits and vegetables, a large category of Mediterranean products, contribute more than 20% to the total value of agricultural production. More than one third of the holdings grows permanent crops (table 6.6). Moreover, the average size of a permanent crop growing farm has half the size of the average holding. Small producers cultivate products with minimal support from the CAP. On the other hand, the cultivation of cereals and of industrial plants is more concentrated in fewer and larger holdings.

Moreover, although three out of five holdings in Greece rear livestock, the composition differs from that of other EU countries. In Greece, sheep and goats account for 55% of total livestock unit as opposed to 11% in EU12, whereas the heavily supported by the CAP bovine animals barely amount to 16.5% as opposed to 52% in EU12.

Institutional inefficiencies constitute a major factor accounting for the changes in the composition of agricultural production, the reduction in the value added per unit of product, and the increase in agricultural trade deficit. It is worth exploring the determinants of these developments, since some restructuring did take place since Greece's accession to the EU, but to the detriment of the sector.

Table 6.6
Land use by agricultural holdings (%), 1993

	Arable land		Permanent pasture & meadows		Permanent crops		Woodland	
	holdings	area	holdings	area	holdings	area	holdings	area
Greece	44	57	7	14	47	28	2	1
Spain	35	43	17	28	37	14	10	16
France	36	61	32	30	14	4	18	5
Italy	34	43	13	21	39	14	14	22
Portugal	35	47	8	19	36	16	21	18
EU12	36	50	21	32	30	8	14	10

Source: Calculated from data presented in *Statistics in Focus,* 1995, Eurostat

Given the technology, major factors affecting rational decision making by producers are, first, output and input prices and, second, the information on current conditions and on likely future developments regarding the markets. Farmers did change production patterns, according to the direction given solely by the price levels of a certain marketing period and by the false perception that this price level was secured as the minimum level in future time. That is producers made rational decisions based on the information they had, but that piece of information was inadequate and disorienting.

No consideration was given to the importance of linking production to the satisfaction of consumer needs or to acknowledging the association between producers' income on one hand and export and economic growth on the other. Farmers were used to viewing their production as the passage to financial support, irrespective of market needs and consumer demands. Financial support became the end goal. There was no other significance attached to production. Thus, producers began to alienate themselves from the production process, from the environment and from the natural resources. Producers received higher production levels as a unique opportunity to raise their income, and they concentrated upon that objective. They stopped safeguarding the environment. Consequently environmental degradation began.

The food chain broke at the producers' link. Farmers stopped thinking that production should be geared to consumers, rather than dumping practices, as final recipients. The price support mechanism, in combination with inadequate information to producers, estranged producers and consumers, removed producers from the eco-system and offered middlemen the opportunity to exploit the emerging gap.[9] Since there was essentially no continuity in the food chain, middlemen were at liberty to set prices at new levels, much higher than those justified by the intermediaries' actual contribution under competitive conditions (Sapounas, 1991; Karabatsou-Pachaki, 1984).

In view of these problems, it is necessary that an appropriate environment be created in order to counteract structural constraints and to reverse the negative implications of globalization and of the CAP operation to the benefit of agricultural and rural economies. Thus, it is imperative that ways be found to improve competitiveness, increase farm holding size, induce specialization in products of high income elasticity and high added value to promote part-time farming, attract young farmers in agriculture and to train them. In this attempt, the role of institutions and of the macroeconomic environment in preserving or removing distortions and rigidities, and in affecting restructuring of the agricultural sector should not be underestimated.

The potential for sustainable endogenous development

Whether effectuating dramatic or moderate changes in agriculture and in rural society, international developments will force the agricultural sector to face increasing competition in the world markets. The option to adjust in the context of conventional agriculture will be obstructed by severe natural and institutional impediments. More importantly, any attempt to exploit economies of size will translate into further mechanization, adoption of capital intensive techniques, higher unemployment in rural areas, and more skewed distribution of income in agriculture. Thus, Greece must explore ways for smoother adaptation of the agricultural and rural sectors to the upcoming transitions.

The intersectoral linkages affecting the allocation of resources necessitate the adoption of an integrated approach toward agricultural development. Especially in Greece, where off-farm income is the main component of total farm income for small family holdings, agricultural prosperity cannot be achieved outside the framework of rural development. Since conditions are constantly changing, rural development cannot be static in nature. Nevertheless, growth should be sustainable, emanating from a system having flexible capacity and ability to adjust successfully to the ever-emerging socio-economic conditions without causing irreversible negative effects in the process. Consequently, the notion of sustainability is directly linked to conservation and preservation of the environment and of the natural resources. Moreover, in order for the system to maintain flexibility through time, it is necessary that economic decisions and performance be well understood and that new situations, needs, and potential to adjust be assessed and evaluated regularly. This dynamic process requires the comprehensive analysis both of individual choices and of institutional constraints.

Given the disparities within the rural and the agricultural sectors and the regional divergence in structure and institutions, rural development should be pursued taking into account differences in local needs and potential. An endogenous approach to rural development would suggest rational use of local resources and exploitation of the local comparative advantage in order to prevent the outmigration of young people, land abandonment and environmental degradation (Merlo and Manente, 1994), while alleviating inequities.

In order for the potential for sustainable endogenous development to be realized, rural areas should create the necessary institutional and structural setting for stimulating regional economy by taking advantage of and using local resources efficiently, while protecting the environment.

Efficient use of resources presupposes rational decision making, full access to information and valuation of the resources at prices reflecting their social benefit or cost. If these conditions were satisfied, then rational decisions would determine the efficient choice of activities to undertake, of the techniques used in pursuing these activities and of the level of products or services to be supplied. The same rules would determine the allocation of resources among alternative uses or activities. Output-output, output-input, and input-input prices would reveal the relative advantage of choosing one product over another, of adopting a specific production technique and of using a certain amount of inputs.

Nowadays productive resources are being misallocated due to the distortions prevailing in the markets of land, capital and labor. Mobility of land is obstructed by heritage law and by the absence of a mechanism for managing land and allocating it into uses. Thus, the opportunity cost of agricultural land is raised, supply of agricultural land is reduced and farm fragmentation is maintained. On the other hand, distortions on the cost of capital has encouraged consumption rather than investment. Finally, rigid institutions impede young farmers from assuming the management of farm holdings and obstruct the creation of alternative job opportunities in rural and urban communities.

The remainder of this section examines potential components of an approach facilitating structural adjustments in agriculture and promoting rural development. Central elements of this approach comprise changes in institutions and in technology, which are the factors determining the costs of production and transactions.

Economic decisions and the creation of rural employment

Just as in every other sector of the economy, the demand for human capital and for labor depends on the Value of Marginal Product of labor (VMP) and is directly related to the price of the last unit produced and sold and to the productivity of the very last utilized unit of labor. On the other hand, the supply of labor is associated with the opportunity cost of labor, with the maximum return that labor could secure if it were channeled into some other use. The value attached to leisure by individuals is also a determinant of labor supply.

By definition, policies which bring about an increase either in the marginal productivity of farm labor or in the agricultural product's price lead to higher demand for labor in agriculture. The marginal productivity of labor could be enhanced by actions and policy alternatives that: 1. expand the use of other means of production (i.e., capital and land); 2. improve the technology applied; and 3. improve the quality of the other production factors by investing in physical capital (e.g., installations, machinery, etc.), in human capital (e.g., education, training, etc.), and in land (e.g., land reclamation projects, etc.). On the other hand, the unit price of the product could be raised as a result of: 1. improvements in the quality of the product; 2. increases in the value added; and 3. adjusting production or marketing services to consumer preferences (e.g., the products' type and characteristics, the production techniques, handling, packaging, storing, processing, and distribution).

Furthermore, labor exits agriculture when employment opportunities outside agriculture exist. Lower supply of labor in agriculture will result from increased marginal productivity of labor in off-farm activities or from a relative rise in the price of non-agricultural products.

Consequently, both the spatial and the intersectoral distributions of labor are determined by factors relating to differences in remuneration and in living conditions. Wage differentials reflect differences in the marginal income of labor and differences in the value attached to additional leisure time. Policy instruments affecting the above factors alter the incentives for spatial or inter-professional mobility of labor.

Structural changes in agricultural labor markets can result from: 1. increases in the demand for labor in agriculture; 2. reductions in the demand for labor in agriculture; 3. increases in the supply of labor in agriculture; and 4. decreases in the supply of labor in agriculture, or from any combination thereof.

Structural changes that increase the demand for labor in agriculture raise employment, wages and income, and facilitate economic progress. This situation, being far from depicting current conditions in Greek agriculture, should constitute policy objective and should be pursued by creating conditions enhancing the marginal productivity of labor in agriculture and resulting in higher prices of agricultural products.

Currently the Greek agricultural sector is characterized by a combination of reduced demand for and increased supply of labor. These structural changes were effected on one hand by technological improvements in agriculture (i.e., to mechanization and substitution of labor by capital) and the pressure from increasing international competition and by the combination of the influx of immigrant workers and higher unemployment in urban areas.

In such cases, the result on income depends on the relative size of the structural changes in the demand and in the supply, as well as on the respective elasticities. If the decrease in demand were larger in absolute terms than the increase on the supply side, then agricultural income would undoubtedly drop, since both wages and employment levels would shrink. If the decrease in demand were smaller than the increase in supply, then the outcome would be dubious, since wages would fall, but employment would rise. A positive outcome could result if the increase in employment, caused by the emerged lower farm wages, were big enough to counteract the drop in unit earnings. Such a situation would be depicted by a need for unskilled labor (elastic demand), which does not correspond to current conditions in agriculture.

Structural changes that reduce the demand for labor in agriculture call for actions promoting increases in demand for labor both within agriculture and in non-agricultural sectors of rural economy. Toward this end, it is necessary that 1. marginal productivity of labor in agriculture and in rural areas increase especially in the production of commodities with falling prices and 2. the composition of output change in favor of products, services and production methods which can guarantee high prices.

The final category of structural change, depending on its origin, can have a desirable effect or not. Lower labor supply in agriculture, if effected by immigration and internal migration, is undesirable, since it leads to depopulation and to the socio-economic decay of rural areas. On the other hand, if it is the result of a rise in employment in non-agricultural sectors it may have positive impact both on agricultural and on rural economies. For a rural economy, the less skilled the labor demanded by the non-agricultural sectors and the more skilled the labor supplied, the higher the increase in non-agricultural income. In agriculture, the more specialized the labor demanded and the more skilled the remaining labor the better for agricultural income. In addition, the positive outcome could be reinforced by actions increasing the demand for labor in agriculture.

The prevention and reversal of depopulation in rural areas urges that employment opportunities increase in the rural sector and the living conditions in rural areas (i.e., quality of work-time and leisure) be improved.

The statements above specify necessary conditions for economic progress in rural areas, as measured by increases in the per capita income and in employment levels. Challenges arising from increased competition and technological improvements cannot be confronted without innovation, novelty, flexibility, and continuous adjustments to changes. Consequently, it is necessary that the efforts to attain goals be entrusted primarily to young farmers. The success of the outcome will depend largely on securing access to information, training, land, and capital for young farmers. On the other hand, the viability of small-sized family farms and the vitality of rural communities require the creation of employment possibilities in the non-agricultural sectors of the rural economy.

Based on the analysis presented above, policy measures intended to enhance development in rural areas should primarily, although not exclusively, aim at (Damianos and Hassapoyannes, 1997):

- Encouraging the installation of well prepared and trained young farmers (under 40 years of age) in agriculture, in order to promote innovations, institutional changes and immediate adjustment to more competitive and market oriented agricultural practices;

- Encouraging young - and especially - single women to engage in labor intensive non-agricultural activities (handicrafts, weaving, on-farm processing, agro-tourism, etc.), in order to increase employment in non-agricultural sectors and to enhance living conditions in rural areas. A common complaint of young men in remote rural regions is that there are no young women to marry, since most of them move to urban or semi-urban centers;

- Encouraging well educated urban inhabitants to move to rural areas, in order to assume specialized, professional positions in non-agricultural rural sectors or to establish non-conventional enterprises, engaging in alternative agricultural methods or in labor intensive non-agricultural activities. This action should be viewed as important mainly for enhancing living conditions in the countryside and secondarily for increasing employment. It constitutes a crucial goal, because it aims not only at securing high quality services in rural areas, but also at eliminating society's negative image for the rural life style, and

- Supporting small family agricultural holdings in engaging in on-farm alternative activities. This action is designed to create and to maintain sustainable agricultural holdings, to enhance living conditions and to preserve the environment.

The need for restructuring the agricultural sector

Greek agriculture should receive the attention it deserves, given the importance of the sector as a major employer, contributor to the country's GDP and trade performance, guardian of cultural heritage and supporter of remote areas. The development of the agricultural sector should constitute a priority goal and prerequisite for economic and social revitalization and for sustainable growth. In this context, the distinctive characteristics of farming as well as the social role of family farming should be acknowledged. Agriculture should be facilitated in pursuing efficient restructuring that would enable the sector to overcome the obstacles raised by international developments and by domestic structural constraints. Toward this end, it is necessary that objectives for the sector be clearly defined and that an instrumental environment to achieving these goals be carefully created.

In determining the objectives, it is helpful to register the sector's needs which were formed as a result of trade liberalization, the policies that promoted intensification of production and broadened income discrepancies, as well as the policies and institutions that effected the detachment of farming from its social identity.

Trade liberalization necessitates sustainable increase in world and domestic market shares, in order to secure long run dynamic growth, job creation and better income levels in agriculture. This need presupposes improved competitiveness of the sector, which entails reduction in the cost of production relative to that in other countries and adjustment of production to the emerging consumer preferences and needs. Nevertheless, particular attention should be paid to measures that reduce the cost of production by taking advantage of economies of scale, since such actions could impair environmental conditions further. Both today's higher ecological awareness and past negative experience warn us about the devastating impact that competition without concern for the environment may have on natural resources. Similarly, the estrangement of farmers from socio-cultural values should be reversed. The cultural identity of rural areas should be reinforced, while the social role of family farming should be acknowledged, and its social dimension should be re-established and ascertained. Furthermore, inequities should be minimized in the agricultural sector, in order to broaden the economic base, raise welfare, and improve social cohesion. This condition should constitute a fundamental principle in agricultural policy design. Finally, it is necessary that agricultural development be seen not as an autonomous process, but as part of and dependent upon the realization of rural development. The next chapter elaborates on this issue and draws attention to the misconceptions raised by the careless use of "improved competitiveness".

Thus, the agricultural sector should be restructured along the following dimensions:

- the economic dimension of developing a competitive and robust sector;
- the ecological dimension of conserving and preserving the environment and the natural resources;
- the welfare dimension of minimizing inequalities in the distribution of income;
- the socio-cultural dimension of providing non-distorting social protection to farmers and of establishing the social role of farming; and
- the political dimension of restructuring institutions so as to raise mobility, to enhance citizen participation and to effect social cohesion.

These objectives have to be pursued within a framework taking into account the problem of income instability attributed to agricultural activities, current institutional constraints determining the mobilization of resources, market distortions effected by the divergence of prices from social values, the rigidity shown by farm organizations and farmers toward adjustment, the role of multi-activity as a major income source for small sized family holdings.

Long term exposure to heavily shielded agriculture and limited access to information cultivated a mentality that accepted price support as an indisputable, incontestable, and irreversible concession to farmers and farm organizations. Production decisions - what, how, and how much to produce - were based solely on price levels for a production period, without considering long term perspectives for domestic and international markets. Producers learned to live in a cocoon, where time effected no change. Farmers misinterpreted support prices as secured rights, which time would, if anything, improve further. Producers stopped viewing themselves and their activities as integral elements of the natural environment and cultural heritage. Farmers stopped viewing the multidimensional contribution of agricultural products and services to society. Having lost their capacity for social contribution, producers accepted the negative image that rural living and agricultural vocation was given by urban centers.

Today, farmers should realize that their services can be expanded beyond the production of goods that are of little interest to consumers. In order for this mentality to change, the state should secure conditions for better access by farmers to information, to programs for infrastructural improvements, and to social justice. In this process, however, producers should take initiatives, exploit the opportunities offered and stimulate rural development by trusting their abilities and local potential. Farmers should be re-introduced to meaningful participation in public life and decision making. Women should undertake roles of equal significance in the management of farm holdings, in the development of diversified activities and in securing supplementary income.

Nevertheless, the mere imposition of regulations, without the endorsement of the directly affected parties, could effect no sustainable change. It is imperative that farmers and farm organizations be informed of the need for restructuring, be persuaded of its wide beneficial effects and be convinced of the feasibility of any implementation plan.

Nowadays, international developments allow more space to market forces for determining prices, raise environmental, nutritional and health concerns, enhance cultural sensitivity. The role of consumers becomes crucial. Quality of life is redefined along these principles. In this setting, uninhibited access to accurate information is essential. Restructuring can take place only if farmers view agriculture as a dynamic activity and not as a static and inert practice. Successful restructuring presupposes continuous access to full information that will permit the formation of realistic expectations by producers regarding the direction of future developments. Efficient restructuring, being a dynamic process just like agricultural activities and the formation of consumer preferences, is not compatible with price support policies and static measures.

While low mobility of production factors (land, quotas, labor) and low investment levels are main determinants of high production costs, insufficient information and training, lack of entrepreneurial mentality, poor education levels are mainly accountable for environmental degradation, loss of market shares due to inability to satisfy the emerging consumer preferences, and deterioration of the quality of life in rural areas. The latter has also been the result of poor incentives in rural areas and of inadequate participation in institutions.

Thus, institutional changes are indispensable for creating an environment conducive to structural improvements in agriculture and in non-agricultural sectors as well. Toward this end, the role that European Community programs can play toward improving infrastructure, raising income and employment levels should not be underestimated.

Notes

[1] Non-uniform prices resulted from the imposition of the Monetary Compensatory Amounts (MCAs) after the fixed monetary system collapsed in August 1971.

[2] Realization of the specialization benefits was halted also by the sub-optimal utilization of labor in agriculture. Relative immobility of farm

labor was due to a combination of higher agricultural prices under CAP and lack of job alternatives in urban areas.

[3] The Central and Eastern European Countries (CEECs) covered are the ten countries which signed the Association or Europe Agreements with the EU between December 1991 and July 1995, namely Poland, the Czech Republic, Slovakia, Hungary, Bulgaria, Romania, Estonia, Latvia, Lithuania, and Slovenia.

[4] The per capita income of the applicant countries equals one third of the Union's average (CEC 1997).

[5] According to the Commission's proposal, spending on the new Objective 1 and 2 regions should be restricted to 35-40% of the Union's population in 2006 rather than to the current 51% (CEC 1997).

[6] World Trade Organization.

[7] According to Sapounas (1991), during periods of high prices for traditional products, the use of arable land was intensive and the application of fertilizers was excessive.

[8] Multi-active family holdings are defined as those having a farm manager younger than 65 years old and spending more than 50% of their total working time on off-farm activities (Damianos et al., 1994).

[9] As Timmer et al. (1983) state, "each party comes to the exchange with different knowledge about the characteristics of the underlying market forces for the item of exchange. Arrow argues that the party with the relatively greater knowledge actually sets the initial price". Lack of transparency and market failures lead to the creation of monopolistic structures and to widening price gaps in the food chain (Karabatsou-Pachaki, 1984).

Bibliography

Buckwell, A., J. Haynes, S. Danidova and A. Kwiecinski (1994), *Feasibility of an Agricultural Strategy to Prepare the Countries of Central and Eastern Europe for EU Accession*, Final Report to the DG-I of the European Commission, December.

Commission of the European Communities (1991), *The Development and the Future of CAP*, Discussion Document (COM91) 100 final, Brussels.

Commission of the European Communities (1995), *Study on Alternative Strategies for the Development of Relations in the Field of Agriculture between the EU and the Associated Countries with a View to Future Accession of these Countries*, Agricultural Strategy Paper, CSE(95)607.

Commission of the European Communities (1997), "Agenda 2000: For a Stronger and Wider Union: A historic opportunity", file retrieved from internet address http://europa.eu.int/comm/agenda2000/rapid/over_en.htm.

Damianos, D., M. Demoussis, C. Kasimis and A. Moisidis (1994), *Pluriactivity in Agriculture and Development Policy in Greece*, Foundation of Mediterranean Studies: Athens (in Greek).

Damianos, D. and K. Hassapoyannes (1997), 'Mediterranean European Countries Strategies in Relation to the Final Act of the Uruguay Round', *Options Mediterraneennes, Serie A, No. 30, CIHEAM.*

Delorme, H. and Clerc, D. (1994), *Un Nouveau Gatt? Les Echanges Mondiaux apres L'Uruguay Round*, Editions Complexe: Paris.

FAO, (1995), *Impact of the Uruguay Round on Agriculture*, FAO: Rome.

Hathaway, D. and Ingco, M. (1995), 'Agricultural Liberalization and the Uruguay Round', World Bank Conference on *the Uruguay Round and the Developing Economies.*

Hill, B. (1984), *The Common Agricultural Policy: Past, Present and Future,* Methuen Series, No. 839, London.

Ingco, M. (1995), 'Agricultural Trade Liberalization in the Uruguay Round, One Step Forward One Step Back?', World Bank Conference on the *Uruguay Round and the Developing Economies.*

Karabatsou-Pachaki, K. (1984), "The Economy and the Agricultural Sector" Report presented at the International Conference organised by the Agricultural Bank of Greece, Athens (in Greek).

Labropoulou, A. N. (1995), The Primary Sector: Evolution and Perspectives for Development, 1994-1999, Reports, No 17, Centre of Planning and Economic Research: Athens (in Greek).

Larsen, A. and J. Hansen (1994), 'Agricultural Support and Structural Development', in "The Economics of the Common Agricultural Policy", *European Economy*, Reports and Studies, No. 5.

Lasaridis, P., E. Kousoulakou and M. Poulakou (1994), *Structural Characteristics and Production Systems of Agricultural Holdings*, Studies of Agricultural Economics, No. 48, Agricultural Bank of Greece: Athens (in Greek).

Mergos, G. I. and D. Psaltopoulos (1996), *The Agricultural Machinery Industry and Agricultural Mechanisation*, Institute of Economic and Industrial Research, Special Sectoral Themes, No. 5, Athens (in Greek).

Merlo, M. and M. Manente (1994), 'Consequences of Common Agricultural Policy for Rural Development and the Environment', in "The Economics of the Common Agricultural Policy", *European Economy*, Reports and Studies, No. 5.

Nallet, H. and A. Van Stolk (1994), *Relations Between the European Union and the Central and Eastern European Countries in Matters Concerning Agriculture and Food Production*, Report to the European Commission, June.

OECD, (1995), *Better Policies for Rural Development*, OECD Document, The Proceedings of the High Level Meeting of the Group of the Council on Rural Development, OECD: Paris.

Papaconstantinou, M. (1991), 'Greek Agriculture and GATT', *To Vima*, Athens.

Sapounas, G. S. (1991), *Agricultural Development: Problems and Prospects*, Studies of Agricultural Economics, No. 42, Agricultural Bank of Greece: Athens (in Greek).

Sapounas, G. and D. Miliakos (1996), *Greek Agriculture in the Post-War Era: The Effects from Acceding the European Union and Lessons for the Future*, Studies of Agricultural Economics, No.51, Agricultural Bank of Greece: Athens (in Greek).

Sarris, A. H. (1994), 'Consequences of the Proposed Common Agricultural Policy Reform for the Southern Part of the European Community', in "The Economics of the Common Agricultural Policy", *European Economy*, Reports and Studies, No. 5.

Sarris, A., Mergos, G. and Sarros, P. (1996), *The Uruguay Round Agreement for International Trade and its Impact on Greek Agriculture*, Foundation of Economic and Industrial Studies: Athens. (in Greek)

Sheehy, S. (1997), 'Towards Free Trade. A European Union Perspective', *Choices*, First Quarter.

Tangermann, S. (1994), 'An Assessment of the U.R.Agreement on Agriculture in OECD', The New World Trading System: Readings, Paris.

Tarditi, S., S. Senior-Nello and J. Marsh (1994), *Agricultural Strategies for the Enlargement of the European Union to the Central and Eastern European Countries*, Final Report to the DG-I of the European Commission, December.

Timmer, P. C., W. P. Falcon, and S. R. Pearson (1983), *Food Policy Analysis*, Johns Hopkins University Press: Baltimore.

Wallden, S. (1994), 'Commercial Relation between Greece and Bulgaria', Presentation at the Conference on the Greek-Bulgarian Economic Relations Towards the Year 2000, Institute of International Economic Relations, Athens (in Greek).

Wallden, S. (1995), *Greece and EU Enlargement towards Central and Eastern Europe*, Hellenic Centre for European Studies, Research Papers, No. 35, September (in Greek).

7 The framework for an integrated strategy for sustainable rural development

Guidelines for national regional and intersectoral policies

The role of agriculture in the Greek economy and in its social structure sets agricultural development as a priority goal. Although the repercussions of international developments already are being felt heavily in agriculture, other sectors are also in need of restructuring. Economic interrelations and implications require that the problems facing Greek agriculture should not be treated separately and independently from those encountered by other sectors. Any attempt to confront the problems presupposes that agricultural adjustments take place within a framework acknowledging intersectoral linkages.

Moreover, agricultural development cannot take place isolated from the macroeconomic environment nor can it overlook particular regional characteristics and special local needs. Agricultural development cannot be realized but as a component of sustainable rural development, supported and forwarded by a set of compatible and stimulating macroeconomic, regional and intersectoral policies and by institutional changes. This, however, requires that agricultural policy should be incorporated into a strategic plan for dynamic and integrated growth of the rural sector. Implementation of an integrated policy for the agricultural and the rural sectors will provide the opportunity to exploit the new comparative advantages of the country towards the revitalization of rural regions and social prosperity.

The planning process of an effective integrated strategy consisting of regional and intersectoral policies presupposes the following:

- awareness of global developments, knowledge of the international framework of action and accurate perception of current tendencies and future trends. These conditions are necessary for identifying the developments that generate new needs, for determining the required course of actions and for taking the path of successful - according to the strategic goals - adjustment to the new conditions. Correct assessment of the developments and reliable evaluation of the emerging needs require unobstructed and continuous access to constantly updated and accurate information. Dynamic restructuring requires that the Greek economy keep abreast of the latest world developments and that it react within the international and the European Union's framework of obligations. In this process of combating distortions and inequalities, Greece should take full advantage of the EU Common Structural Policy provisions and of the accompanying CAP reform structural measures;
- a well delineated and distinct picture of: 1. the current social and economic structure of the target groups; 2. the various socio-economic inter-relations and inter-linkages; and 3. the specific local characteristics and needs within the national territory. This condition is necessary for evading poor selection of target groups and consequently of policy measures that could produce undesired results. Fulfillment of this condition sets the framework of restructuring and assists in specifying the criterion for determining the appropriate measures (incentives and disincentives) of national structural policies for achieving the goals, without introducing new distortions and additional inequalities. In this context, it is imperative that the importance of pluriactivity in forming total income of small sized family holdings be fully acknowledged and taken into account for policy making; and
- clearly defined end goals and target groups. In order for rural areas and for the agricultural sector to develop in a sustainable way, the goals should comprise minimization not only of inefficiencies but also of inequities. The Greek economy needs to grow in a way that disparities in the distribution of income between rural and urban areas, between the rural and the agricultural sector and within both the agricultural and the rural sectors be minimized. Harmonious development of the entire national territory is essential for expansion of the income base, stimulation of income growth and employment, promotion of social cohesion and political stability.

Under those conditions, an effective strategy in advancing sustainable rural development along the economic, ecological, socio-cultural and political dimensions presented in the previous chapter should be based on the following principles:

- improved competitiveness;
- efficient restructuring and operation of institutions;
- valuation of resources, products and services so as to reflect social costs or benefits; and
- decentralization and adaptation of policies to special local characteristics and needs. Any policy that disregards these requisites may not only prove ineffective, but it may also introduce new problems.

Improved competitiveness

The problems created by the gradual demolition of protectionism and by globalization can be summarized in the intensifying efforts required by the country to maintain and to increase its world market share in a sustainable way. A larger value share for Greek products in world markets would raise foreign exchange revenues and would improve the trade balance, thus facilitating the contraction of deficits, the reduction of borrowing needs and the drop of interest rate levels. Such a development would stimulate private investment, raise national income, improve infrastructure and create new jobs. The Greek economy should restructure itself in order to escape from increasing trade deficits, high interest rates, low investment levels and high unemployment. More specifically, it is imperative that the economy in general and the agricultural sector in particular become competitive.

Nevertheless, competition is not - as usually presented - a self-justifying goal. Such careless interpretation may lead first, to the adoption and implementation of ineffective and harmful measures to the economy and second, to the exclusion from consideration in the decision making process of several crucial factors for economic development. Some elaboration on the term is due before closer examination of the problems.

Often times, the term "competition" is used to characterize the situation where a country's products and services fare well in world markets. In this context, input subsidies and export restitutions could be thought of as a means to raise competitiveness of Greek products, since these measures would lower the price of the products supplied in the world markets and would result in increased consumption by foreign consumers.

Nevertheless, this interpretation ignores that the benefits accrued from competitiveness relate to that attribute's capacity to restrict the needs for financing deficits, to encourage investment and to stimulate economic growth. Policies that result in higher exports to the detriment of national budget deficits do nothing but reduce opportunities for investment and in the long run, raise unemployment. Policy measures that effect artificial increases in exports to the detriment of the national economy should not be perceived as enhancing competitiveness. Since a country's competitiveness in supplying a specific product is gauged by the relative unit cost of domestic production with respect to the unit cost of production in foreign countries, a meaningful interpretation of improved competitiveness could be provided by the reduction of the actual unit cost of domestic production relative to the unit cost of production abroad.

It is necessary that the Greek agricultural sector undergo major restructuring to become more competitive. Such an objective necessitates simultaneous, coordinated and joined efforts toward changing the composition of products and services supplied for domestic and foreign consumption, adding special attributes to goods and services offered and effecting reductions in relative costs of production. In this process, it is imperative that services be given at least as much attention as products and that any actions undertaken to improve competitiveness be comprehensive. Moreover, restructuring, should take place in ways that would guarantee intersectoral accordance and cooperation for improving the competitiveness of interrelated sectors, protecting the environment and the natural resources, keeping cultural heritage and values alive, safeguarding the social role of agriculture, family farming and rural communities.

To epitomize, market shares, in an increasingly competitive environment can rise either by effecting lower actual unit costs of production than in the competing countries or by providing a slightly differentiated product, embodying distinct attributes and services, reflected in higher value added and unit price. Attempts to increase market share by means of implementing distorting policies (e.g., production and export subsidies), at best will defer confrontation with the problems of marginalization and at the worst will aggravate the situation. Therefore, there should be no illusion that satisfaction of frequently advanced demands for subsidization of the production cost (e.g., petroleum) will improve the competitiveness of Greek agriculture. The only results it could have would be to ameliorate farmers' income in the short run. Nevertheless, this would probably induce consumption and would put upward pressure on inflation rates. Furthermore, the use of subsidized inputs would increase. Consequently, such measures would misallocate resources and would have a negative impact on the environment due to intensification of production, on employment due to substitution of capital for labor and on equity, since larger enterprises would benefit more from having the capacity to exploit economies of scale.

Institutions and organizations

It is crucial that the definitive role that institutions and farm organizations have in the performance of agricultural and rural economy be acknowledged and be exploited in the process of restructuring toward sustainable rural development. The importance of this issue requires a closer look and an introduction to what is meant by "institutions" as well as to the relation between institutions and farm organizations.

Institutions comprise humanly devised formal rules and informal codes of behavior that shape human interaction by structuring social, economic, or political incentives in order to reduce the uncertainty involved in the completed production process. When a product or a service is supplied into the market, it embodies the input costs (i.e., the costs of land, labor, capital) that accrued from: 1. transforming resources into output; 2. transforming the physical attributes of a good in form, time and space; and 3. transacting. Transaction costs include the costs of defining, protecting and enforcing the property rights to goods, that is the right to use, to derive income from the use of, to exclude and to exchange goods. In a perfectly competitive environment, where complete information is available at no cost to all agents and where property rights are well defined, institutions do not matter. But in the real world of imperfect markets, institutions arise in order to facilitate human interactions and thus to minimize, together with technology, the costs of exchange and production (North, 1996).

The ability of institutions to minimize the uncertainty and therefore the cost of risk involved in human interactions determines economic performance. A complex institutional framework of formal rules, informal constraints and enforcement combined makes low-cost transactions possible by providing credible environments for making agreements and for enforcing them. Thus, unclear definition of institutional constraints, poor incentive structure and loose enforcement of the rules make institutions ineffective.

While institutions determine the opportunities in a society, organizations develop and evolve in consequence of the institutional framework with the purpose of achieving selected objectives. In this process, organizations form and reform strategies, use skills, experiment and learn by doing, which in turn affects the institutional framework. Since organizations can be political bodies (e.g., political parties, city councils, regulatory agencies), economic bodies (e.g., family farms, cooperatives), social bodies and educational bodies (e.g., schools, vocational training centers) (North, 1996), it follows that an integrated strategy for sustainable rural development cannot but address the ineffectiveness of institutions and pursue their restructuring toward efficiency.

This is served best by introducing proper motivation for both operating and enforcing the rule structure and by creating an environment conducive to forming accurate perceptions (Ostrom, 1996). Thus, institutional adjustment requires right incentive structure for both public sector agents and users, open and transparent governments, the empowerment of stakeholders and effective decentralization of public services, administrative and political competition, and the involvement and empowerment of indigenous institutions (Klitgaard, 1996).

An efficient incentive structure should direct agents toward productive rather than redistributive activities, should effect the creation of competitive solutions rather than monopolies and should expand rather than restrict opportunities. The institutional framework that prevailed in the Greek agricultural sector fit perfectly North's description (1996) of the incentives that led to "persistent failure" in taking the path to economic growth. It effected high transaction costs in economic markets, due to incomplete and poorly processed information and to structural constraints impeding the mobility of productive resources. In addition, it seldom induced investment in education that could increase productivity. As a consequence to redistributive incentives, agricultural and rural organizations (firms, cooperatives, etc.) became efficient in making the society even more unproductive.

In addition, absence of secure and well defined land property or renting rights in Greece increased uncertainty, raised costs of transacting and resulted in the adoption of technologies that used little fixed capital and did not entail long-term agreements. With land renting conditions of informal agreements of uncertain and short term tenure, it was unreasonable to expect high investment levels and long term horizons in agriculture. The observed substitution of crop for livestock production resulted partly from this fact.

Institutions in Greece contributed largely to the stagnation of agricultural and rural economy. The achievement of sustainable rural development and of economic growth necessitates not only structural but also institutional adjustment. Toward this end, it is imperative that institutional restructuring should complement structural adjustment. Yet it should be borne in mind first, that institutional changes occur in an incremental and marginal fashion and second, that a mixture of informal norms, rules and enforcement should be restructured.

In order for agricultural and rural restructuring to become efficient, it is imperative that the underlying Greek institutions change so that incentives and in turn organizations be redirected along lines that increase productivity (North, 1996). The tax structure, inheritance law, land policies should change so as to raise mobility of resources and to shape the kinds of skills and knowledge that pay off, since the direction of knowledge and skills is the decisive factor for long-run development.

Organizations should invest in knowledge that increases the productivity of the physical or human capital or that improves the perceptive ability of entrepreneurs. Farm and rural organizations should disengage themselves from getting the government to introduce distorting policies and to adopt make-work practices. In addition, informal rules have to change. Informal constraints are important especially in raising labor productivity, since the attitude of labor affects not only the quantity but also the quality of the output that it produces. Ill-formed perceptions have to be identified and combated mainly by facilitating the access to full and accurate information (Ostrom, 1996). The formation of conventions and forms of organization that encourage conscious work participation and cooperation with an ideological commitment build a "morale [that] is a substitute at the margin for investing in more monitoring" (North, 1996).

> Rules that encourage the development and utilization of tacit knowledge and therefore creative entrepreneurial talent [are] important for efficient organization. Obviously competition, decentralized decision making and well specified contracts of property rights as well as bankruptcy laws are crucial to effective organization. ... It is essential to have rules that eliminate not only failed economic organization but failed political organization as well. (North, 1996)

Again the selection of incentives is crucial. As stated by Ostrom (1996) "causes of past poor performance and the sources of improved performance, lie in the incentives facing providers". Many infrastructure projects have been unsustainable because "perverse incentives (face) participants in the design, finance, construction, operation, maintenance and use of facilities". Thus it is important that the factors affecting incentives of the participants in activities be well understood. Moreover, both designers and enforcement agents of projects should have intrinsic or extrinsic incentives attached to the project, so that self-enforcement would be possible.

Social valuation of resources, of products supplied and of services performed

Social valuation is also linked to the efficient operation of institutions. Institutions alter the terms of exchange by setting a mechanism of incentives. Furthermore, factors that affect the terms of trade in turn have an impact on incentives and gradually alter the institutions.

Any changes that could result in prices reflecting social - rather than private - profitability would introduce a whole new structure of incentives in the operation of institutions and of organizations. Such changes could result from sectoral, intersectoral or macroeconomic policy decisions. But why should resources, products or services rendered reflect social values? Why should prices and therefore the incomes accruing to each factor of production, convey signals of relative factor availability and reflect social opportunity costs?

Timmer, Falcon and Pearson (1983) argue that policy interventions that distort the factor prices by subsidizing interest rates or by imposing minimum wage rates lead to misallocation of resources, to unequal access to factors of production and to wider inequalities in the distribution of income. In addition, distorted prices may lead to production decisions that are inappropriate and inefficient for the local environment. Such policies, although most often chosen to assist more disadvantaged farm holdings, end up benefiting the efficient entrepreneurs and to worsen the position of the initially targeted group of farmers.

Prices that do not reflect social opportunity costs introduce both inefficiencies and inequities. This is usually the result of distorting policies, of limited access to information, access to inadequate or incomplete information, false perceptions or externalities.[1] Economic efficiency requires efficient price formation, which in turn presupposes:

- full and accurate information being equally accessible by all producers (sellers) and consumers (buyers);
- accurate and real expectations about future conditions. "Dynamic price formation ... integrates information about future crops and alternative supplies, demand pressures and storage costs to allocate the supplies in hand to future time periods.... The temporal pattern of prices established or the price expectations formed, signal producers, consumers and the suppliers of storage as to the opportunity costs of their production, consumption and storage decisions" (Timmer et al., 1983); and
- reconciliation of the differences between private and social profitability, for farmers make their decisions on the basis of the market signals they actually perceive, not those used by analysts in a planning agency.

The contribution of the pre-reform CAP to the shift of resources to low valued crop production and to environmental decay exemplifies the negative impact that divergence between actual prices and social values had in Greece. Partly due to the incomplete and inaccessible, to most small farmers, information partly due to distorting policies and externalities, the use of pesticides and chemical fertilizers increased dramatically and resulted in soil degradation and in depletion of valuable resources, such as water.

This development was common to all EU member states. The CAP reform of 1992 acknowledged the threat on the environment, included a new objective for the protection of the environment and switched to new policy measures that decoupled support from production and provided incentives for environmentally friendly methods of production. Moreover, the reformed CAP promoted the role of farming in countryside stewardship, without however assigning any prices to the qualitative attributes of rural areas (e.g., landscape, cultural heritage, etc.) or specifying any means for remunerating this type of services.

Although the 1992 reform did change the CAP orientation, many more things need to be done. Several issues should be discussed further and daring actions should be taken. Even the new objectives and measures cannot guarantee efficiency and equity without social valuation. A structure of proper incentives for promoting the operation and the enforcement of environmentally friendly practices should be carefully designed. "Rather than telling farmers how much pesticide to use and trying to enforce the rule, pesticide prices must be set to reflect full social costs. Incentives to plant crops that cause less soil erosion or replanting schemes to stabilize barren lands may be more effective than police action to prevent farmers from cultivating hillsides to villagers from poaching firewood from public lands" (Timmer et al., 1983).

Valuation according to social profitability could effect not only more efficient use of consumable inputs, but also the placement of investments in a social context (Timmer et al., 1983). This is a particularly important issue, especially under current conditions of making the best use of the second Community Support Framework and of making plans for implementing the third one. Several times, especially under the political pressure to show large absorption rates of Community funds, investments are undertaken based on the impression that their size makes rather than on the efficiency with which capital is used. Yet, "the productivity of an investment in a society is as important for growth and the alleviation of poverty as the investment itself" (Timmer et al., 1983).

Even in the current CAP, which is supposedly more sensitive to environmental issues, nothing guarantees that the adopted production controls lead to efficient use of chemicals. Since the yields that were used to calculate the maximum production levels embodied intensive use of resources, nothing ensures that the post-reform CAP leads to efficient use of pesticides. Furthermore, although distortions are gradually removed by promoting the transparency of measures and by bringing EU prices closer to world levels, nothing prevents large scale enterprises that practice capital intensive agriculture from creating inefficiencies by imposing externalities on small scale organic farms. Is the promotion of organic products better served by the currently granted subsidies per unit of land devoted to organic farming or rather by taxing the use of hazardous to the environment inputs in conventional agriculture?

Figure 7.1 illustrates the problems that arise when private profitability differs from social profitability, as happens in the case of externalities (Just et al., 1982). The MPC curve depicts the marginal private cost of producing Q or the supply cost of Q in the presence of externalities. The MEC curve depicts the marginal external costs imposed externally on all affected agents. At P_0, the competitive solution in the presence of externalities results in the production of quantity Q_0. This outcome, stemming from equating prices to marginal private costs, is much larger than the quantity Q_1 that would be produced at the same price level P_0, if production decisions accounted for the marginal costs effected externally in addition to the marginal private costs. It can be seen that the total external costs that accrued from the production of Q_0 units are equal to the area $k+h$, that is the area below the MEC curve or, equivalently, the area between the MSC (marginal social cost) and the MPC curves, from the beginning of the axes up to the quantity level of Q_0.

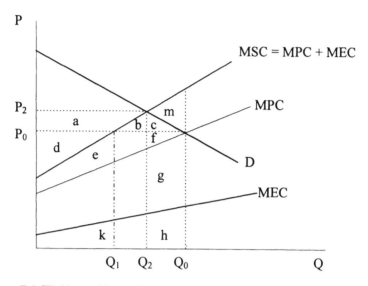

Figure 7.1: Welfare effects of an externality

The socially optimal solution, however, is obtained at price level P_2 and quantity Q_2, where market price equals marginal social cost. First, it can be observed, that in the presence of externalities less output will be produced and it will be sold at higher prices than in the competitive solution, where the impact of external costs was not taken into account. At Q_2, there is no incentive to society to expand output, since that would imply social cost larger than the social utility per additional unit of output sold (MSC > D). Furthermore, at Q_2, there is no incentive to society to restrict output, since consumers would be willing to more than compensate for a decrease in social costs, in order to be able to consume that additional quantity (MSC < D).

In moving from the competitive to the socially optimal solution, consumers lose area $a + b + c$ as a result of both the increase in price (i.e., area $a + b$) and the decrease in the quantity sold (i.e., area c); producers gain area $a + b$ from selling at a higher price and lose area f from decreasing the quantity produced and sold in the markets. Finally, external agents gain area h from the effected reduction in external costs. Thus, the net social gains equal $- c - f + h$ or equivalently area m, since by construction area $h =$ area $c + f + m$. Three things are worth mentioning. First, the impact on producers is ambiguous,

216

since inelastic demand may imply gains for them. Second, according to this measurement of social values, "social optimality does not necessarily imply that externalities should be restricted to zero" (Just et al., 1982). Third, society unambiguously gains.

The last conclusion triggers the consideration of another issue. If in the presence of externalities the switch from private to social profitability makes society better off, then what type of policies should be introduced in order to achieve social optimality?

In general, three types of measures are employed in the above context: 1. Pigouvian taxes or subsidies; 2. standards (pollution quotas); and 3. assignment of property rights (Just et al., 1982).

"A Pigouvian tax is a tax that imposes the external cost of pollution on the generator of that pollution". In figure 7.2, the competitive solution is at P_0 and Q_0. A socially optimal solution results from equating marginal social benefit (MSB) to the marginal social, rather than private, cost. Consequently, production falls to Q_1 and price rises to P_1. In order to achieve social optimality, however, a tax equal to the marginal external cost associated with the desirable output is imposed on the producers of the pollution. Thus, at production level Q_1 the Pigouvian tax imposed on polluters is $P_1 - P_2$. Market (consumer) price is now equal to P_1, whereas producers receive P_2 after the tax. Thus, producers lose area $e + f + g$, consumers lose area $a + b + c$, government gains area $a + b + e + f$ in tax revenues and external agents gain area $c + d + g$ as a reduction in external costs. Therefore, society on the net gains area d.

Alternatively, the social optimum can be attained by means of standards (pollution quotas) to control externalities. This case, adopted by the reformed CAP for major product categories, entails the direct, restriction of production to level Q_1 (figure 7.2) by the government. As a consequence of the drop in production, the price that both producers and consumers face rises to P_1. Therefore, the distribution of income changes from the Pigouvian tax case, although the new welfare gain to society is the same (area d). More specifically, producers gain area $a + b - g$, consumers lose area $a + b + c$, and external agents gain area $c + d + g$. That is, this scheme differs from the previous one only in that producers gain, whereas no revenues accrue to the government. Actually, under the standards scheme, consumers transfer area $a + b$ to producers, since they pay higher prices than in the absence of the control. Under the tax scheme, however, both consumers and producers transfer surplus to the government, equal to areas $a + b$ and $e + f$ respectively.

Finally, social optimality in the presence of externalities can be obtained by assigning property rights. More specifically, according to this method either law L assigns polluters the right to pollute, or law L_n assigns pollutees the right not to be polluted (right to pollution reduction). Essentially, this method "encourages a market for the externality", since "if a polluter clearly owns the right to pollute, the pollutee may be willing to pay the polluter to either reduce or cease pollution. Alternatively, if the pollutee owns the right to no pollution, a potential polluter may buy the right to pollute from the pollutee" (Just et al., 1982).

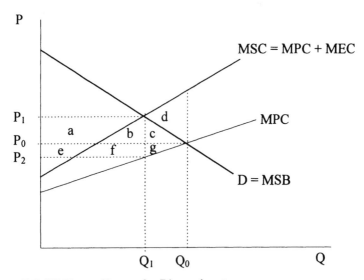

Figure 7.2: Welfare effects of a Pigouvian tax

Nevertheless, depending on the assignment of property rights, the effect on income distribution varies widely. In the case of L laws, the polluter can trade his right to pollute with transfers by the pollutees equal to $(P_1 - P_2) (Q_0 - Q_1)$ (figure 7.2). Thus, the pollutees transfer income to the polluters and pollution is reduced from levels corresponding to output Q_0 to levels corresponding to output Q_1. In the case of L_n laws, the pollutees can sell their right not to be polluted in exchange for transfers by the polluters equal to $(P_1 - P_2) Q_1$. Thus, pollutees receive income from the polluters and pollution corresponding to output Q_1 is generated. In general, polluters have a higher by $(P_1 - P_2) Q_0$ income when they own the right to pollute and the pollutees have a higher by $(P_1 - P_2) Q_0$ income when they own the right to pollution reduction. The assignment of property rights presupposes that a few parties are involved as polluters and pollutees, that none of the parties involved may exercise market power, and that the source of pollution can be identified.

The validity of the above analysis requires that several general points should be stressed:

- In the previous examples, pollution is considered to be in fixed proportion to production. With pollution abatement equipment or techniques, however, pollution can be reduced without a similar reduction in output. In this case, the Pigouvian tax or the standards mechanism must be imposed directly on pollution rather than on output. "Otherwise, production is discouraged and the appropriate incentive for use or installation of pollution abatement equipment is not conveyed to the polluter" (Just et al., 1982). This case questions the effectiveness of the production controls imposed by the reformed CAP as a means to curtail the use of environmentally hazardous chemicals in the production process. This is even more important in the case of Greece, where excessive use of consumables is ordinary. Since the reference quantity for quotas is based on performance that did make heavy use of fertilizers and other pollutants, it is likely that the quotas will restrain production but without leading to environmentally conscious use of inputs. Therefore, if it is desirable to restrict the use of chemicals in the EU, a tax should be imposed directly upon the pollutants. As it stands now, the policy certifies that the primary concern of the EU was to alleviate the budgetary pressures and to reduce the structural surpluses rather than to reduce environmental pollution and waste of resources.[2]

- "The optimal imposition of standards or quotas may require a substantial amount of information on individual cost curves, which, if not obtained, can suggest a preference for use of taxes in controlling externalities." (Just et al., 1982). If the individual cost structures are ignored and the same level of quotas, Q_q, is applied indifferently to all farmers, then at producer price P_1 and quota level Q_q,[3] some farms would be induced to operate above their optimum level Q_1^*, (e.g., firm 1 in figure 7.3), while others would be forced to operate below their optimum level Q_2^*. This situation is inefficient, since it leaves space for overall improvement, should firm 1 reduce production, while firm 2 expanded. For the use of standards to be preferred, two conditions should hold: first, monopoly rather than competition should prevail and the quota level should correspond to the socially optimum production point. In this case both taxes and assignment of property rights lead to second best solution, because of the existence of two distortions (externalities and monopoly).

- In order for the socially optimal level of a pollution tax or of standards to be determined, the damage function of the externally affected individuals (external agents) should be known. This requires a lot of information, which presupposes at least a market of property rights of farmers. In the case of Greece, the Register of Farmers could be used to estimate damage and control cost functions (Just et al., 1982).

In contrast to the above mentioned policies for controlling externalities, the subsidies granted to organic farming entail budgetary costs and the danger of perpetuating the mentality of relying on government support. Most importantly, they do not restrain environmental damage and degradation, since organic farming subsidies are intended to facilitate environmentally friendly practices, but without putting intensive conventional farming to any disadvantage or under any control.

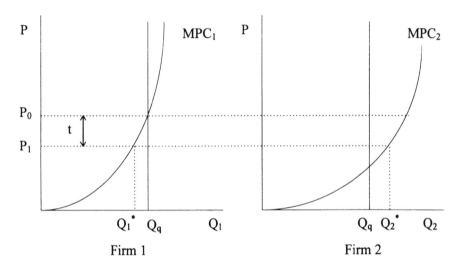

Figure 7.3: **The effectiveness of quotas and taxes in controlling externalities**

Policy adaptation to special local characteristics and needs

The importance of the local conditions in policy making and policy implementation is directly related to the performance of institutions as described earlier in this section. Better knowledge of local conditions, special characteristics and needs will assist in identifying and forming better incentives for promoting self-enforcing agreements in human interaction for designing policies, for implementing and enforcing them.

Agreements and contracts promote stability in human interaction and are self-enforcing when they provide no incentive for any of the contracted parties to break the agreement. North (1996) claims that small communities form a social environment conducive to low cost human interactions, because of the knowledge that parties have about each other and the repetitions in dealings. As the repetition of a problem allows for better processing and reduction of transaction costs, local agents can be much better observers of local repeated problems and thus better processors. Therefore transaction costs are minimized when information is obtained locally and processed by local agents. On the other hand, governance of projects by local participants is likely to be more effective, since local agents "craft their own rules - almost invisible to outsiders - which frequently offset the perverse incentives they face in their particular physical and cultural setting" (Ostrom, 1996).

The necessity for local participation in policy making stems from the dependence of the effectiveness of institutional constraints on "the motivation of the players (their utility function), the complexity of the environment and the ability of the players to decipher and order the environment (measurement and enforcement)" (North, 1996). Since institutions also consist of informal rules, better knowledge of local natural conditions and of social conventions is necessary for specifying the appropriate motivation structure and for minimizing the costs of transaction and transformation. Technology should also be adapted to specific local characteristics and needs in order for transformation costs to be minimized.

Ostrom (1996) argues that it is more important that national governments create an environment conducive to the development by local public entrepreneurs of "a wide variety of efficiency-enhancing solutions to local collective-action problems" rather than "attempting to plan and build local infrastructure throughout a country". National governments should build institutions that would enhance the capacity of local public entrepreneurs "to organize, mobilize resources, and invest in public facilities, ... [and to] provide fair and low-cost conflict resolution mechanisms, methods of achieving public accountability and good information about the conditions of natural and constructed resource systems".

With regard to the successful operation of organizations, Ostrom (1996) also states that project plans should reflect accurately local conditions and needs. Moreover she argues that incentive systems of operation and maintenance divisions either should be drawn and enforced from the directly affected participants or should reward non-local enforcement agents for drawing on local knowledge and working directly with farmers. In general, the incentives faced by the designers and the enforcement agents should be closely aligned to the incentives of the farmers who will be directly affected by the project. This way, the performance of the system is linked to that of the officials and both types of agents have incentives to design and to enforce efficient operation of the system.

Finally, "the efficient functioning of the marketing systems is particularly sensitive to local, cultural, and social conditions and especially to the local availability of resources ... [including] managerial, administrative, and entrepreneurial resources" (Timmer et al., 1983).

Although all indications argue for decentralization, it should be stressed that the mere shift of governance responsibilities and other public activities from urban centers to small rural communities will not result in efficiency. In order for decentralization to be effective in promoting sustainable rural development, it should be supplemented by a mechanism that aligns the incentives of public agents with those of private entrepreneurs and of policy designers with those of the directly affected rural citizens.

> Adaptive efficiency ... [concerned with the kinds of rules that shape the way an economy evolves through time ... the willingness of a society to acquire knowledge and learning, to induce innovation, to undertake risk and creative activity of all sorts, as well as to resolve problems and bottlenecks of the society through time] ... provides the incentives to encourage the development of decentralized decision making processes that will allow societies to maximize the efforts required to explore alternative ways of solving problems (North, 1996).

To summarize, an effective integrated policy approach presupposes "reading [of] long-run international market trends and using the signals to measure the efficiency of domestic price policy initiatives. It also includes careful attention to the domestic food marketing sector, ... because of the marketing sector's role in generating and signaling prices". It is for this reason that policies for rural development should "include other prices important in rural decision making, especially wage rates, interest rates and foreign exchange rates" (Timmer et al., 1983).

Central elements of the restructuring process towards sustainable rural development

The restructuring process towards sustainable rural development should be viewed as a multi-axes pursuit along the guidelines presented in the previous section. Thus, rural development should be pursued on the basis of a carefully planned, comprehensive strategy aimed at building a strong economic base for rural zones and at improving living conditions. The first and to a large extent the second target will be achieved by means of:

- reducing inequalities in the distribution of agricultural income;
- developing the agricultural and the non-agricultural sectors in rural areas; and
- establishing connections and relations with new domestic (local and non-local) and foreign markets.

In addition, improvement of living conditions has a social dimension, which requires public attention. Nevertheless, indiscriminate financial support magnifies inequalities in income distribution, effects loss of competitiveness and burdens society as a whole. Social agricultural policy should guarantee basic services to everybody, but should be selective in providing additional social support to persons or areas in need. It follows that social welfare, taxation, technical support and other systems should be re-examined under this prism. Similarly, mountainous and remote areas also should be treated favorably.

Equitable distribution of income presupposes the creation of a strong labor market and easy and equal access by small agricultural holdings to improved means of production, job opportunities, information, incentives and to new markets. The attainment of sustainable development depends on the capacity of rural areas to create real, self-sustained job opportunities, especially for young people, keeping these regions from remaining depopulated. Finally, the access of new or the expansion to currently accessed domestic or foreign markets boosts up development and strengthens the operation of the labor market. The means of achieving sustainable rural development are interrelated and should be pursued in a coordinated manner.

These goals require the adoption of new technology, an infrastructural and an institutional environment conducive to mobilization as well as to efficient use of resources and output and compatible, sound and well planned macroeconomic and sectoral policies that take into account local conditions, needs and advantages.

Job creation policies

The surplus labor emerging from agriculture or from urban areas cannot be absorbed by the current rural and agricultural economic structures. It is imperative that the problem of unemployment in rural areas be dealt with in the context of a global, integrated policy for rural development. Macro-economic and intersectoral policies combined, taking into account the local characteristics and needs of each region should create a fostering environment for encouraging the development of new rural activities and economic diversification (OECD, 1995).

Labor market policies that enhance mobility and incorporate sufficient flexibility to deal with rural conditions are vital for improving employment opportunities. The development of a labor market both in the agricultural and in the non-agricultural sectors of rural areas requires increases in the demand for labor by means of an increase of the marginal productivity of labor and a shift of production in favor of high value-added products or services. In order for these conditions to be attained, policies should aim at promoting investments that would result mainly in new and better suited technology to local conditions, improvements in the quality of production factors, such as human capital, physical capital and land. More intensive use of capital and land would also raise the marginal productivity of labor. Nevertheless, the adoption of capital intensive techniques could displace labor and increase unemployment, while increases in the availability of land would require substantial restructuring of the institutional framework and investments on irrigation and land reclamation.

Also important are investments that would raise the quality and the value-added of products and services and would adjust them to domestic and foreign consumer preferences. Thus, farmers' involvement in economic activities should expand beyond the production process, to the secondary and tertiary sectors. This implies engagement in all marketing stages (storage, processing, distribution, etc.) and in trade, which can be feasible through the participation of farmers, especially of small-scale farm holders into more efficient large scale organizations and cooperatives.

Consequently, the establishment and the adoption of on-farm alternative activities provide good opportunities for promoting sustainable rural development. Such activities require restructuring of the holding and rearrangement of the means of production (land, capital, labor and managerial skills). The objective is to produce on-farm a combination of new products and services, including those of non-agricultural nature and to take advantage of new forms of organization of production (e.g., cooperatives). The pattern of on-farm production secures that the members of family farms remain employed by their holdings and receive the value added of their own labor by engaging in on-farm, agricultural or non-agricultural activities, whereas in the usually observed case of multiple job holding or pluriactivity, the value added by off-farm family labor is extracted by the employer.

In general, there are three types of alternative activities that could be adopted by farmers:

- production of crop and livestock products which are new to a specific area, non-conventional in nature or in their intended use;
- adoption of labor intensive techniques and production methods. This type of activity includes development of extensive production systems, reduction in the use of capital relative to labor and especially of all industrial means of production (fertilizers, chemicals, etc.), production of organic, ecological products and in general environmentally friendly management of agricultural holdings; and
- production and on-farm marketing of labor intensive, agricultural and non-agricultural products and services (e.g., agro-tourism, agro-manufacturing, on-farm processing of agricultural products by the farm

family members). Any activity which increases the value added -whether by processing or by on-farm sales- or offers a service to the benefit of society (e.g., countryside stewardship) belongs in this category.

In any case, public sectoral policies should stimulate local entrepreneurship by the private sector or by cooperatives for new products, new markets - including the development of niche markets- and better business practices. Increasing the number of entrepreneurs, improving the business climate and developing new mechanisms for cooperation between private sector agents will contribute to generating employment opportunities. Nevertheless, it is important to encourage regional diversification along with local specialization (OECD, 1995).

Sectoral or regional incentives should be accompanied by the appropriate macro prices, fiscal control and monetary growth that would foster investment decisions conducive to the creation of productive jobs (Timmer et al., 1983). Inflation and real interest rates should be kept at low levels. Moreover, lower production costs should be pursued by non-distorting measures, such as improvements in infrastructure and marketing networks, reductions in land cost and increases in the productivity of human capital.

Given the geopolitical and cultural heterogeneity of the rural space in Greece, policies should be adjusted to the specific characteristics of each region. Scientific and research institutions should specify policy proposals, taking into account all facts at the national, regional and local levels. The effectiveness of public organizations and institutions at any level in providing consultation and technological know-how is a decisive factor for the development of alternative enterprises.

It is imperative that the government create a favorable environment for taking advantage of new and emerging opportunities, which sometimes differ from those in the past. Policies should create an environment that provides access to information about international trends to rural communities. This would allow the agricultural sector to keep abreast of changes that will enable them to acquire flexibility and adjust successfully. Besides the effective dissemination of information, high quality consulting and technical services should be available for improving the quality of decision making. To this end, it is imperative that administrational restructuring take place, that extension services be well organized and that substantial technical assistance be given to farmers. Agronomists should be relieved from bureaucratic tasks and should be assigned to field work.

It is the state's responsibility to establish an environment that would stimulate the mobility of production factors and improve the accessibility of available resources. Such a framework requires mainly institutional changes, such as the following:

- redistribution of land. If properly implemented, this measure may lead to the absorption of excess labor by securing economies of scale. Creation of jobs in agriculture requires drastic restructuring of tenure relationships and rural asset ownership. Moreover, taxing non-agricultural uses of agricultural land would lead to more rational allocation of the means of production, because it would trigger an increase in the supply of land for agricultural use. As a result, labor, capital productivity and employment would increase and income inequalities would be reduced. Appropriate tax measures should be introduced also to other input uses to ensure better signaling of social costs by market prices. Above all, a new institutional setting should facilitate the access of young farmers to land;
- liberalization of capital movements and enhancement of capital accessibility. It is imperative that capital be allocated to productive uses. Capital movements should be liberalized, so as to direct savings toward productive private investment. Then, public investment should complement private investment (CEC, 1986). Thus, it is critical that low income farmers in remote areas be given equal access to incentives for using improved propagation material and industrial inputs. Similarly, it is equally meaningful to supplement large irrigation projects by wide, small-scale irrigation networks that would serve smaller farms as well. In general, irrigation and other networks should be developed in an integrated way, serving agricultural holdings, irrespective of their holding and income size. Measures that lead to distortions should be avoided. Instead of an indiscriminate subsidization of the cost of capital, incentives of fiscal or other nature should be provided for promoting investments that would reduce the cost of production; and
- investment in human capital. Employment policies should counteract the obstacles in effectiveness raised by distance and low population density. Improved education and training are critical for the development of rural areas. It is imperative that educational programs, technical and vocational training be promoted for the diffusion of knowledge, especially of the know-how relating to alternative farming systems. Applied knowledge is equally important to pure knowledge. Farmer to farmer training programs are essential, since farmers of an area or of areas with similar characteristics can convey their valuable experience regarding applied

productive and efficient systems (Ostrom, 1996). In addition, adequate incentives should be given in order to encourage early retirement and the establishment of young people in the countryside and in agriculture. More specifically:

- young farmers should be guaranteed better living conditions and access to capital, know-how, information and land. Farm management practices need to be updated and adjusted to current needs. Special emphasis should be given to offering information about new possibilities available to young farmers, professional training and incentives for installation. Full exploitation of measures related to early retirement and Community Initiatives, such as LEADER, are also necessary. Structural Funds available for less favored areas should finance activities which guarantee an integrated and spherical approach to rural development. The existing common tools for the implementation of the Common Structural Policy could be supplemented by additional measures, such as training before establishing the farm business, encouragement of installing alternative farming activities, housing loans and especially to young couples who decide to start up an agricultural enterprise, targeted tax exemptions, priority to the redistribution of production quotas of land, special grants for the establishment of new commercial enterprises by young farmers;
- young female farmers should also be treated favorably and be given priority in incentives promoting agro-manufacturing, agro-tourism and processing on the farm. National measures for facilitating the installation of this category should include relaxed tax measures, as well as advertising and information programs regarding tourist services in rural areas;
- well educated or especially skilled young urban dwellers should be encouraged to migrate to rural areas. National management of structural policies should secure access to training programs for the promotion of alternative farm activities, to the provisions of the Common Structural Policy and to Community Initiatives; and
- finally, small agricultural holdings should be induced to take advantage of the accompanying measures of the CAP reform for the promotion of differentiated and diversified agricultural activities. Nevertheless, the applicable policy measures should be adjusted to the specific conditions facing farms and regions.

National measures

On the national level measures should be taken so as to examine the potential contribution of the following to the development process:

- establishing priority rural zones as a first step to identifying target groups and deciding upon the nature of preferential treatment of financial charges by zone;
- promoting new activities for better management of rural space by removing some types of limitations and constraints and by alleviating the financial burden of some entrepreneurial categories. The encouragement of adopting on-farm activities presupposes a thorough understanding of the pluriactivity phenomenon and of the particularities that rural inhabitants face. (The Farmers' Register could be used to apply direct social assistance selectively to target groups in need.);
- improving the effectiveness of public institutions and organizations and incorporating local agents in decision making. The incentives of national government officials should change so that that their work enhances rather than replaces the efforts of local officials and citizens;
- ensuring coordination and cooperation between different levels of government and improving rural governance. Central government should support both regional and local leadership, in a way that innovation is encouraged at lower levels and that different actors work independently toward common goals. Successful and innovative coordination and cooperation among the public, private and voluntary sectors could be a key factor to the integration of rural economies in the global market place.

Environmental protection

Special care should be given to the preservation of the environment and to natural and cultural resources. Intensive farming and livestock production have caused environmental pollution and depletion of water resources. National and regional programs for water management should be implemented. Moreover, cultivation techniques and investments should be radically reformed so as to promote environmentally friendly practices.

Policies that acknowledge the positive relationship between well maintained, small rural communities and the quality of the natural environment between the upkeep of rural and traditional lifestyles and culture should be forwarded. Rural areas with their landscape, life-styles and indigenous cultures, unique products, monuments and history are critical in preserving national identity. Provision of rural amenities should be a goal, viewed not as an excuse for transfers to inhabitants of rural areas, but as a means to create national wealth and income. The effectiveness of policies with such attitudes, however, presupposes proper quantification and therefore remuneration of the services provided by rural amenities and their measurement so as to reflect social benefits as accurately as possible.

For a successful environmental strategy, the institutional framework should change in order to support not only the implementation, but also the enforcement of environmental policy. Targets should be clearly specified, goals should be well defined and priorities should be specifically set. Independent management agents should be established at national, regional and local levels. Products and services should be assigned prices and taxes that reflect social costs and benefits. It should be reminded that negative side-effects of development practices and particularly environmental pollution, stem from market distortions that incur from the divergence between market prices and social valuation of the product or of the service marketed.

Several arguments against organic farming, for example, state that the higher cost of organic products makes them non-affordable for the budget of an average consumer, thus suggesting the cost advantage offered by conventional agriculture to consumers. This type of argument, however, is correct[4] only in depicting accounting costs in product prices. This logic collapses however, when the economic decisions by consumers and producers are taken based on input and output prices reflecting the *social cost* to society. Thus, if the social cost imposed on the health of current consumers, of future generations and on the environmental quality were reflected in the price of the often overutilized chemical fertilizers used in conventional agricultural, their use would be lower and the unit price of the products obtained this way would be higher. Then, a cost comparison would be in favor of organic products. Consequently, a system of assigning social values to products used and produced should be developed. "Green" taxes should be imposed on the purchase of non-organic agricultural inputs, rather than subsidizing organic farming in order to protect markets from being distorted again. Furthermore, farmers offering services toward conservation of the environment, natural resources, cultural heritage should be remunerated according to the benefit accrued to society.

Land management and taxation

The current situation of high rents, high land values and high transaction costs places a burden on the cost of production, makes investment a less attractive option, impedes the attraction of young and dynamic farmers in agriculture. Well defined and enforced property rights are essential for reducing transaction costs. Consequently institutions that would facilitate the management and the transactions of land (e.g., a land bank) could enhance the factor's mobilization, could promote long term land tenure contracts and provide incentives for long term planning and for substantial investment in agriculture.

Agricultural land should be bestowed by ownership or by use to farmers in ways that reduce fragmentation, increase farm size and create economically viable enterprises. This objective, however, presupposes the classification of available land according to its uses, while it is sensitive to the specification of the term "farmer".

A land policy should aim at:

- the protection of agricultural land;
- the rational use of agricultural and forest land;
- improvements in the structure of agricultural enterprises, in order to take advantage of the size with respect to the production activities performed;
- the organization of production activities on a voluntary co-operative basis in order for the effective use of agricultural and forest land to be expanded by collective operation; and
- the allocation of public property land into production activities (e.g., by yielding its use to cooperatives or local administration).

In order to alleviate the skewed distribution of land, prevent the cost of agricultural land from rising and raise land mobility, a tax system should be formed so that resources would be better allocated. More specifically, a land taxation scheme could have the form of: 1. a tax on real estate; 2. a tax on non-utilized agricultural land; or 3. a tax on transferring agricultural land to non-agricultural uses (Balfoussias et al., 1988).

Information and Education

Since "market knowledge is market power" (Timmer et al., 1983), both information and education systems are required in order to strengthen the performance of farmers and cooperatives. Information systems could provide basic knowledge, information and education could enable farmers to decipher this piece of information and to process it in the best possible way. Nevertheless, society should direct investment on education to productive knowledge. Therefore, primary rather than higher education should be given priority in investment, since it is primary education which has a much higher social rate of return (North, 1996).

It is proposed that the establishment of electronic network centers for the information and education of rural population be carefully examined. Such networks could be designed to allow rural citizens and farmers to have simple and direct access to up to date information on domestic and international matters and developments and communication with administrative centers, research institutes, universities and other organizations. Educational subjects and training could also be promoted through such a network.

Cooperatives

Cooperatives can be determining factors to rural revitalization by supporting family type enterprises and by establishing social justice. The size of cooperatives gives them the economic capacity to take advantage of economies of scale and rational organization. It secures better terms of trade to cooperative members. Moreover, it ensures better allocation of resources and alleviates or even neutralizes the impact of price and production fluctuations on farm income.

Restructured cooperatives should develop activities leading to increases in the value added of products and services and should assume a central role in educating, training and preparing farmers, especially younger ones, to adopt new methods of production and innovating activities. Flexible management is needed to appreciate and effect the expansion of cooperative activities into the marketing, processing and trade sectors, as well as to the development of the services sector in rural areas. Nevertheless, cooperatives should pursue infrastructural investments that enhance productivity rather than technologically impressive undertakings of poor value to local conditions.

Agricultural cooperatives should be restructured and their role should be upgraded so as to offer meaningful services to their members, to agricultural households, to rural areas and to society as a whole. Cooperatives can provide their members with sufficient bargaining power in input and output markets, increase the mobility of capital, lower information costs and spread risk. Nevertheless, this capacity presupposes the motivation of members, their ability to decipher the environment and their capacity to enforce the rules that secure an economically rational, democratic and cohesive action. Voluntary participation should be based on the belief by candidate members that the net benefits that would accrue to them being members of the cooperative would be larger than those they would receive, if they operated independently. The stronger the incentives by participants for effective governance, management and enforcement the better the institutional rules created by them will be. Thus, farmers should be exposed to the benefits that can be extracted from establishing and enforcing rules for the efficient involvement of cooperatives in various activities.

Moreover, farmers and cooperatives alike should assume their responsibilities. Nowadays, cooperatives cannot stand by their members because they are financially weak. Unless the cooperatives' financial obligations are "strongly tied to a responsibility for repayment over time",

231

membership would aim at "rent seeking [rather] than at [undertaking] productive investment activities" (Ostrom, 1996).

Many of the problems facing cooperatives in Greece today stem from their past choices to operate detached from the real market conditions for a long period of time, ignoring international conditions and developments in world markets. Cooperatives should be instrumental in securing a strong position in international markets for Greek agriculture, in revitalizing the rural sector and in effecting prosperity for society as a whole.

Toward an increased, mutually beneficial, economic cooperation with Central Eastern Europe, the Balkans and the Black Sea Region

In the framework of developing a comprehensive and integrated strategy for sustainable rural and agricultural development, Greece should also take advantage of the opportunities that the constantly expanding new markets of Central and Eastern Europe and the wider region offer. Nevertheless, in order for both the CEECs and Greece to benefit from the development of economic cooperation, socio-economic cohesion should be enhanced and political and economic stability should be established and secured in the region. The accession of the CEECs to the EU is a critical factor contributing to this end. Taking advantage of EU membership, Greece's assistance in securing the CEECs adjustment to EU regulations and in facilitating their access to the EU markets can prove valuable. Thus, Greece can proceed by:

- helping CEEC competent institutions to familiarize themselves with EU legislation on veterinary and phytosanitary issues and with harmonizing the grading and classification system;
- assisting in the organization of a legal and institutional framework for the development of a farm credit system, the establishment of land market, the restructuring of co-operatives;
- informing the CEECs on the operation of product quality certification and on inspection mechanisms;
- transferring technology, exchanging experts and organizing educational and training programs, as well as seminars on the operation of the CAP;
- taking advantage of national and Community programs such as PHARE, TACIS, INTERREG II, for providing technical assistance and for supporting cross-border investment;
- participating in the structured dialogue procedures on Community policies;
- mediating in the promotion of mutual interests for designing new Community Initiatives and in forming new EU positions; and

- promoting the role of the drachma (the Greek currency) as an intermediary currency in transactions.

Besides the assistance to adjustment, both trade and investment could benefit the CEECs by contributing to the expansion of the income base, by reducing inequalities in income distribution, by developing more efficient production techniques and by creating job opportunities. Higher and more equally distributed income would alter consumer preferences to the benefit of products with higher income elasticities (e.g., meat, fish, olive oil, dairy products). In conjunction to the latest views on nutrition, an expanded income base could delineate a particularly favorable future for white meat, fish, olive oil and fresh fruit and vegetables, which constitute current or potential Greek exports.

On the investment side, it is essential that Greek enterprises invest in the secondary sector, especially by setting up a well organized marketing network (storage, distribution, grading, etc.) that would guarantee the continuous and regular supply of markets, as well as into the processing industry and into advertising.

Promotion of trade between Greece and the CEECs

As far as agricultural trade is concerned, Greece and the CEECs do not pose a threat to each other. The CEECs tend to specialize in products of minor exporting importance to Greece (livestock, feedstuff, vegetable oil, sugar), whereas important Greek products do not face competition from the CEECs (e.g., olive oil, cotton, rice, some fish). Nevertheless, feta cheese, wine, tobacco varieties and wheat are faced with current and potential competition. As far as the fresh fruits and vegetables category is concerned, the CEECs produce different products and in different seasons than Greece. Thus, there is no competition in exports of olives, fresh cucumbers, citrus, apricots, peaches, watermelons, currants and grapes. In the long run some problems may arise in compotes and tomato paste (Wallden, 1995).

CEEC accession will raise the volume of products with EU structural surpluses (e.g. dairy products and sugar), which, with the exception of cereals, are not of direct interest to Greece. On the contrary, Greece could benefit from an increase in agricultural exports and from imports at lower prices. In general, with the exception of sheep and goat meat, CEEC agriculture is complementary to Greek agriculture rather than competitive. Furthermore, the CEECs could potentially absorb the surpluses of Greek agricultural products. Thus, Greece should take advantage of the new opportunities and enter the trade and marketing channels dynamically, even in the form of joint ventures.

233

Trade between Greece and the CEECs increased due to a rise in Greek exports of agricultural products and especially of fresh and processed fruits and vegetables and tobacco. On the other hand, Greek imports of traditional CEEC products, such as live animals, meat and dairy products, feedstuff and primary farm inputs dropped or remained constant. Nevertheless, the value both of exports and imports remains low. This is important especially since EU exports of close substitutes to CEEC products rose significantly more than exports of the complementary Mediterranean products. Consequently there is leeway for Greece to take advantage of the potential for expanded trade flows.

Imports from the CEECs to Greece represent a small and relatively constant share in total Greek imports. Dynamic CEEC products, being potential competitors of Greek products, do not exist. Nevertheless, Greece should not rest on the current situation of weak or non-existent competition from the CEECs for the following reasons:

- after the changes in the terms of trade which will accrue from trade liberalization, accession to the EU and the future CAP reform, the CEEC basis of comparative advantage may change; and
- Greece would have to compete not only in the CEECs but also in the EU, especially in labor intensive products and to a lesser extent in agricultural products.

Consequently, in the long run, Greece should not rely on current advantages from proximity, the presence of Greek immigrants in the CEECs and the CEECs' situation at present. In order to maintain its market share, Greece would have to improve its competitiveness and to restructure the composition of Greek exports so as to consist of high quality and high value products, such as masticha-lentiscus, truffles and foie gras, other new products of non-conventional nature or use, name of origin products and organic products. Moreover, Greece should adopt aggressive, modern promotion methods, confront poor marketing organization and non-systematic exports, improve product verification and quality control, promote product-specific advertising in the CEECs, develop counter trade in order to overcome convertibility problems and establish joint venture trading companies by attracting EU capital.

Investment choices

Apart from trade, investment is another sector that promotes development. Investment decisions are based not only on expected effectiveness, but also on the need or the usefulness of presence in a foreign market (Rizopoulos, 1994). The CEECs provide an environment of low labor costs, availability of primary

234

inputs, but also higher risk for investors and poor legal and institutional infrastructure. Large and powerful firms would prefer to take advantage of export opportunities and to avoid the risk ensuing from moving to the CEECs. On the contrary, firms with small market share, facing fierce competition domestically or internationally, would be willing to take the risk of expanding to the CEECs in order to have a chance for maintaining or improving their position. This has been practiced particularly by small Greek firms, which have the advantage of proximity, common cultural characteristics and traditional relations with most of the CEECs and of the flexibility in operating under weak infrastructural conditions.

Under the pressure of globalization, Greek enterprises should become more competitive by diversifying, expanding their size and entering dynamically into a large number of markets (Rizopoulos, 1994). Moreover, larger firms might take advantage of advanced technology, productivity and their dynamic presence in domestic markets for developing new products, exporting them and eventually expanding activities in foreign countries.

The appropriateness of the investment to be undertaken in the CEECs varies with the expanding firm's characteristics and with the situation in the host country. Investment options include: 1. the creation of a subsidiary enterprise; 2. the buy-out of a CEEC state enterprise during the privatization process; and 3. the establishment of a joint venture with local or third partners.

The first option offers competitive advantage over labor intensive activities, special technology, strong reputation-brand name, where the alternative of restructuring involves too high a cost. The alternative of buying-out a CEEC state enterprise during the privatization process is appropriate, if the expanding firm cannot take advantage of low labor costs or of advanced or special technology and if there is need for a well operating distribution system. Finally, joint ventures provide an appropriate solution to firms that are in need of functioning distribution channels abroad and of bypassing trade barriers. These needs arise when the firms wishing to expand abroad have to confront problems of land property rights and of the legislative framework for direct foreign investment. Such a firm would need to have the willingness to share its technological know-how. This option exhibits low entrepreneurial risk and requires relatively low level of capital investment (Maroudas, 1994; Rizopoulos, 1994).

In order to become more competitive and increase their market share, Greek firms should undertake the type of investment that is more suitable to current conditions. Moreover, it is critical for Greece to switch direct investment from the tertiary to the secondary sector, especially for activities in which Greece has comparative advantage (Maroudas, 1994).

In conclusion, in order for Greece to develop a dynamic and powerful presence in the CEECs and in the EU, it should not only take advantage of trade and investment opportunities, but it should also take initiatives for the smooth harmonization and convergence of CEECs' agriculture to the EU agricultural sector.

Notes

[1] "An externality is defined as the case where an action of one economic agent (i.e., producer or consumer) affects the utility or production possibilities of another in a way that is not reflected in the marketplace" (Just et al., 1982). Thus externalities constitute a case of market failure. This category of actions includes the pollution of soil, water, air, and agricultural products with chemical fertilizers and pesticides that several farmers apply. The products of intensive farming are sold at lower prices that do not reflect the social cost of the damage imposed on the environment and on human and animal health.

[2] Nevertheless, it should be pointed out, that in other countries, where farmers acted more rationally, the use of chemicals was intensive but appropriate for raising output to the desirable level with respect to the market prices. Thus, it is likely that, according to the law of diminishing returns, even a small reduction in output at very high levels of production will release much higher levels of input from use. The difference in the case of Greece might be that, in this country, the use of chemicals was exorbitant even for the desirable level of output. Thus, reduction in output might not affect the usage of inputs.

[3] The quota level Q_q, corresponds to the imposition of a tax t (supposedly socially optimal) upon the market price P_0, so that producers receive product price P_1.

[4] Many supporters of organic farming claim that the cost of organic farming drops and even becomes smaller than that of conventional farming in certain cases, after several years pass and the natural equilibrium in symbiotic organisms is restored.

Bibliography

Balfoussias, A., M. Sakellis, and A. Vorlow (1988), *The System of Taxing and Transferring Resources into the Agricultural Sector and Reform Proposals*, Planning Issues, No. 37, Centre for Planning and Economic Research: Athens. (in Greek)

Commission of the European Communities (CEC) (1986), *Reduction of unemployment in a more dynamic European economy: for an effective implementation of the Community co-operative strategy*, European Economy, Annual Economic Report 1986-87, No. 30, November.

Just, R. E., D. L. Hueth and A. Schmitz (1982), *Applied Welfare Economics and Public Policy*, Prentice Hall: Englewood Cliffs, N.J.

Klitgaard, R. (1996), "Comment on 'Incentives, Rules of the Game, and Development', by Elinor Ostrom," in Bruno, M. and B. Pleskovic (eds.), *Annual World Bank Conference on Development Economics 1995*, The World Bank: Washington, D.C.

Maroudas, L. (1994), 'Western Direct Investment in Bulgaria', presentation at the Conference on the Greek-Bulgarian Economic Relations Towards the Year 2000, Institute of International Economic Relations, Athens (in Greek).

North, D. C. (1996), *Institutions, Institutional Change and Economic Performance*, Cambridge University Press.

OECD (1995), *Better Policies for Rural Development*, OECD Documents, The Proceedings of the High-Level Meeting of the Group of the Council on Rural Development.

Ostrom, E. (1996), 'Incentives, Rules of the Game, and Development,' in Bruno, M. and B. Pleskovic (eds.), *Annual World Bank Conference on Development Economics 1995*, The World Bank: Washington, D.C.

Rizopoulos, G. (1994), 'The Pressure of Competition and the Strategies of Western Entreprises in Bulgaria', presentation at the Conference on the Greek-Bulgarian Economic Relations Towards the Year 2000, Institute of International Economic Relations, Athens. (in Greek)

Timmer, P. C., W. P. Falcon, and S. R. Pearson (1983), *Food Policy Analysis*, A World Bank Publication, Johns Hopkins University Press: Baltimore.

Wallden, S. (1995), *Greece and EU Enlargement Towards Central and Eastern Europe*, Hellenic Centre for European Studies, Research Papers, No. 35, September.

Conclusions

In the previous chapters it has been established that although the implementation of national and common policy measures in the agricultural sector has led to increases in productivity and supply, such policies have been less effective with respect to improving consumers' welfare and the general standard of living in rural areas and to safeguarding the viability of rural communities and the quality of the environment.

Greek agriculture is currently faced with a rapidly changing international environment and the challenges of the 21st century. The viability, competitiveness and sustainability of the agricultural sector, as well as the maintenance of a vigorous social fabric in rural areas is at stake, whereas consumers' claims for quality and safe food products has to be guaranteed.

The Common Agricultural Policy is about to be reformed and readjusted, according to the guidelines expressed in "Agenda 2000", placing emphasis on rural development. Price supports will be further reduced and the ensued loss of income will be partly counterbalanced by income compensations. Changes in market and trade policies will be accompanied by the continued and reinforced implementation of rural policy instruments.

The increasing importance of environmental needs and countryside stewardship and management will offer new development opportunities for rural Greece. In this context, emphasis and priority should be given to infrastructural improvements affecting the productivity of agricultural activities. Similarly, non-agricultural activities such as protection of rural heritage, tourism, restoration of the landscape should be promoted in rural areas.

Restructuring of the agricultural sector necessitates the in depth understanding of both the problems and the proposed solutions. Many times, actions of good intent fail because the means are not clearly distinguished from the final goal. Then, problems should be reexamined and solutions should be respecified. An attempt should be made to identify the root of the problem and the candidate solution. What is essentially the reason for supporting the agricultural sector? What is the reason for requiring improved competitiveness? Why is investment considered beneficial? Why is an educational and training system for farmers pursued? Why are cooperatives, young farmers or a land policy necessary for development? Why is high CAP support sought?

These questions highlight the fact that protection and conservation of agriculture are not an end goal, demanding the support of the sector at any cost, to the detriment of other economic activities. The real objectives consist of the following:

- maintenance and attraction of young and active people in rural areas, in order to assist in the economic and cultural revitalization of the rural society.
- employment and income opportunities for rural inhabitants and development of an autonomous economic base in rural areas, regardless of the nature of the sectors providing these opportunities.
- reduction of conflicts and imbalances in the distribution of income and social services.
- protection, rehabilitation and upgrading of the environment and the natural resources. Evaluation of the environmental conditions taking into account the impact on future generations.

In this context, agriculture constitutes one of the sectors that could assist in achieving these objectives. Consequently, the agricultural sector should be dealt with by acknowledging it as a means for promoting the sustainable development of rural regions and not as the final objective. The agricultural sector should become a contributor to the development of the national economy.

In order to achieve these goals, market and institutional conditions should encourage private initiative by farms, cooperatives, or professional organizations. In other words, market conditions should be altered so as to guarantee the mobility of production factors and to facilitate the access of farmers to them. At the same time, old distortions should be removed and the introduction of new ones should be discouraged. It is necessary that:

- equal access by farmers to upgraded production factors, markets, information, employment opportunities, technology, and know-how be established;

- continuous, adequate and reliable information and intelligence be supplied to all farmers, in order for rational expectations to be formed and rational decisions based on long run perspectives to be taken; and
- prices of inputs, products and services be linked to the social and not to the accounting cost.

If these conditions are satisfied, then the private initiative will develop the appropriate activities and actions for the local situation.

Presently the role of the government is strongly interventional. The government, however, should try to limit itself to the role of a facilitator. In this context, it is the government's responsibility to:

- create and guarantee conditions favorable for undertaking and developing entrepreneurial initiative at individual and cooperative levels. The development of infrastructure and information and training systems, the improved operation of administration systems, the supply of technical assistance and extension services should be the government's responsibility;
- supply an integrated system of social protection;
- emit signals that are non-conflicting, compatible and consistent with the goals to the agricultural population. It is also imperative that the macroeconomic policy, expressed by the way that inflation, taxation, interest rates and other measures are confronted, should enforce rather than neutralize the agricultural or rural policies.

The government should minimize the distortions by restructuring the institutional framework so as to create the appropriate conditions for encouraging private initiatives and for transmitting clear signals to farmers and rural population. The mentality has to change and several points should be stressed.

- The term "motive" has to be detached from the notion of financial support and grants and has to be linked to the financial interests stemming from any entrepreneurial activity. Institutional restructuring has to be in compliance with that logic. Legislative actions should not interfere with the specification of activities, conditions, and operational terms of individuals, cooperatives or professional organizations. The undertaking or the abandonment of private initiative should be directly related to the expectations for entrepreneurial benefit or damage, respectively. Entrepreneurial incentives should guarantee and safeguard rational management and efficient operation of the institutions. A specific and detailed institutional framework, centrally designed, is likely to create new problems rather than resolving old ones.

- Competition is not enhanced, but rather violated, by the conventional reductions in the cost of production that input and output supports effect. The benefits from competition stem from the restriction of budget deficits, the encouragement of investment and economic development. Policies that raise exports at the expense of public deficits increase unemployment in the long-run. Increases in the market share should be effected by real reductions in the cost of production or by increases in the value added. The latter could be achieved by the supply of better quality products or of products that are differentiated with respect to their attributes.
- Investments should be pursued insofar as they raise productivity of production factors. Investments should not be evaluated based on their budget. In addition, policy makers should keep in mind that increases in productivity could result in redistribution of income among the factors of production and essentially among the owners of the factors. If, for example, the goal is to increase the income of farm labor, then the choice of investment should be based on the increase in the productivity of labor that could be achieved without displacing labor to the benefit of capital. Unfortunately, the evolution of agricultural investment in Greece shows that a large percentage of it took place in the irrational mechanization of farms.
- The educational system should prepare farmers to confront any relevant development with long term planning. On the other hand, information should aim at providing the farmer with the necessary means to decodify the signals from national and international markets and at assisting him in making rational decisions. Limiting the role of education to that of a supplier of credentials will have no desirable effect.
- Cooperatives should operate on the basis of economic performance only. Cooperatives should exist as long as farmers appreciate that the benefits they can extract as members of an organization with a stronger negotiating market position are larger that those they could receive acting as individuals. Government intervention is undesirable, as long as the government is unaffected by the the operation of the cooperatives.
- A land policy should aim at facilitating the access of farmers to land. Low cost of land is definitely desirable, albeit under the condition that it is not achieved by introducing new distortions. The value of land increased because agricultural land was in short supply and it reflected agricultural income capitalized in it. Also, other uses began to compete successfully with agriculture, partly due to the uncertainty that years of high inflation and unstable macroeconomic environment caused. Its cost can be reduced by providing disincentives for the further shrinkage of agricultural land, for example by taxing the non-agricultural uses of agricultural land. There is a

danger that inappropriate government intervention in the land market could result in a further decrease in the supply of agricultural land in the market.

- The price support system implemented within the CAP framework favored, despite the original claims, the concentration of Community supports into a small number of large enterprises which were in a position to expand the already large scale production of supported products. At the same time, intensification of production benefited these enterprises even more by allowing their costs to fall, while they were degrading the environment and exhausting natural resources, without undertaking any cost.

- The CAP reform (1992) decoupled support from production, but maintained the inequalities, by providing compensatory payments on the basis of yields that were achieved during the period of intensive cultivation. At the same time, the CAP, in "attempting" to promote environmentally friendly practices, introduces new distortions without eliminating the old ones. Providing support for environmentally friendly rather than taxing -according to social costs- unfriendly practices does not guarantee environmental protection. It encourages the irrational expansion of practices that are currently considered desirable and it does not limit detrimental practices to the environment.

- The government's position toward the anticipated new CAP reform and structural policies under the "Agenda 2000" negotiations should be formed considering the total benefits, even if a few farmer groups will be hurt. While the final European Commission proposals are unknown, Greece should realize that according to the currently known scenario, large price reductions concern cereals, milk and beef. Consequently, Greece benefits, in total, since it is a large importer and a relatively small producer of these livestock products. In addition, the feeding cost of producing livestock will be reduced as the cost of grain will drop. Agricultural trade balance will improve and income transfers from Greece to the suppliers of imports will be reduced. If the position that agricultural development should be pursued as a necessary but not sufficient condition for sustainable rural development, then Greece could benefit greatly by the European Community's proposal for the new Structural Policy.

Greek agriculture should take advantage of the opportunities associated with the anticipated enlargement toward the countries of Central and Eastern Europe. The prospect of opportunities for the agribusiness sector, however, presupposes the enhancement of the Greek agricultural sector's competitive position in domestic and foreign markets. The country's comparative advantage, especially in the production of specialized quality products, still has to be fully exploited and conditions in processing and marketing of agricultural products should be further improved and adjusted to new demands.

At the time when more drastic CAP and WTO reforms are imminent, efforts should be made by all parties involved and at every level to increase the effectiveness and efficiency of agricultural and structural policies for the countryside, by tailoring them to the specific needs and characteristics of Greek agriculture and rural society.

Printed and bound by CPI Group (UK) Ltd, Croydon, CR0 4YY

21/10/2024

01777088-0007